Moritz Röger

Optically Powered Highly Energy-efficient Sensor Networks

Karlsruhe Series in Photonics & Communications, Vol. 10
Edited by Profs. J. Leuthold, W. Freude and C. Koos

Karlsruhe Institute of Technology (KIT)
Institute of Photonics and Quantum Electronics (IPQ)
Germany

Optically Powered Highly Energy-efficient Sensor Networks

by
Moritz Röger

KIT Scientific Publishing

Dissertation, Karlsruher Institut für Technologie (KIT)
Fakultät für Elektrotechnik und Informationstechnik
Tag der mündlichen Prüfung: 20. Juli 2012
Referenten: Prof. Dr.-Ing. Dr. h. c. Wolfgang Freude
Prof. Dr. sc. nat. Jürg Leuthold
Prof. Dr.-Ing. Jürgen Becker

Impressum

Karlsruher Institut für Technologie (KIT)
KIT Scientific Publishing
Straße am Forum 2
D-76131 Karlsruhe
www.ksp.kit.edu

KIT – Universität des Landes Baden-Württemberg und
nationales Forschungszentrum in der Helmholtz-Gemeinschaft

KIT Scientific Publishing 2013
Print on Demand

ISSN 1865-1100
ISBN 978-3-86644-972-5

Optically Powered Highly Energy-efficient Sensor Networks

Zur Erlangung des akademischen Grades eines

DOKTOR-INGENIEURS

der Fakultät für Elektrotechnik und Informationstechnik
des Karlsruher Instituts für Technologie

genehmigte

DISSERTATION

von

Dipl.-Phys. Moritz Röger

aus

Stuttgart – Bad Cannstatt

Tag der mündlichen Prüfung:	20.07.2012
Hauptreferent:	Prof. Dr.-Ing. Dr. h. c. Wolfgang Freude
Korreferenten:	Prof. Dr. sc. nat. Jürg Leuthold
	Prof. Dr.-Ing. Jürgen Becker

Table of Contents

Zusammenfassung ..... **i**

Achievements of the Present Work **5**

Summary **7**

1 Sensor Devices and Networks **1**
- 1.1 Fields of Application for Sensors 1
- 1.2 Optically Powered Devices 2
- 1.3 Networks of Optically Powered Devices 4
  - 1.3.1 Media Access Control (MAC) Protocols 6

2 Converting Light in Electrical Current **13**

3 Low-Energy Media Access Control (LE-MAC) Protocol **21**

4 Applications with Low Power and Low Bandwidth Demand **31**
- 4.1 Advanced Monitor Device in Access Networks 31
  - 4.1.1 Access Network Scenarios 39
  - 4.1.2 Access Network Measurements 43
- 4.2 Sensor Devices in Fiber Networks 51
  - 4.2.1 Optically Powered Sensor Network 52
  - 4.2.2 Low-power Electronic Circuits 53
  - 4.2.3 Applications with Low Power and Low Bandwidth 61

5 Applications with High Power and High Bandwidth Demand **63**
- 5.1 Video Transmission System without Control Channel 63
- 5.2 Video Transmission System with Control Channel 68

Summary and Future Work **77**

Appendix A **79**
- A.1. Lower Limit for the Saturation Current of a Diode 79
- A.2. Impedance Matching 82
- A.3. Demarcation Device 84
- A.4. Optically Powered Sensor Node 87
- A.5. Constants 90

Glossary **91**

References **95**

Acknowledgements (German) **103**

List of Publications **105**

List of Figures

Fig. 1.1 Optically powered sensor node with base station ... 3
Fig. 1.2 Optically powered sensor network with its base station and nodes shown in Fig. 1.1 ... 5
Fig. 1.3 Network consisting of a central controller and several nodes connected via a shared medium ... 6
Fig. 1.4 Categories for medium sharing techniques ... 7
Fig. 1.5 Timing charts of different MAC protocols for WSNs (a) Preamble sampling (b) Preamble sampling with recipient information (c) Preamble sampling with recipient and receiving time information ... 10
Fig. 2.1 Equivalent circuit of an illuminated photodiode ... 14
Fig. 2.2 Current voltage characteristic of a photodiode for different illumination powers ... 15
Fig. 2.3 Current and power density of an illuminated photodiode as function of output voltage ... 16
Fig. 2.4 Theoretical efficiency of an optimum cell ... 17
Fig. 2.5 The incoming optical power is equally distributed to *N* photodiodes where the optical power is converted to electrical power ... 18
Fig. 2.6 Current and power density for arrays of 1, 2, 4, and 8 photodiodes as function of output voltage ... 19
Fig. 3.1 Photonic network with optically powered subscribers ... 22
Fig. 3.2 Timing chart schematic for low-energy medium access control (LE-MAC) protocol ... 25
Fig. 3.3 Network of ultralow duty cycle subscribers ... 26
Fig. 3.4 Measured timing chart for low-energy medium access control (LE-MAC) protocol ... 28
Fig. 4.1 Demarcation devices in an FTT*x* network ... 31
Fig. 4.2 Block diagram of the demarcation devices ... 32
Fig. 4.3 Receiver circuit consisting of transimpedance amplifier, inverting voltage amplifiers, and Schmitt trigger circuit ... 33
Fig. 4.4 Transfer function of the transimpedance amplifier, measured and simulated ... 33
Fig. 4.5 Measurements on the performance of the receiver circuit at 80 nW total input power ... 34
Fig. 4.6 PRBS of length 2^{31}-1 in the downstream signal at different optical input powers ... 35
Fig. 4.7 Influence of the PRBS length on the control signal ... 35
Fig. 4.8 Driver circuits for two different kinds of lasers ... 36

Fig. 4.9 Photographs of the printed circuit boards for a demonstrator of a demarcation device ... 37
Fig. 4.10 Photographs of the demarcation device demonstrator ... 37
Fig. 4.11 Current-voltage characteristics of 4 to 8 photodiodes in an optically parallel and electrically serial connection ... 38
Fig. 4.12 Single photodiode open circuit voltage in dependence of optical power and saturation current ... 39
Fig. 4.13 TDM-PON with optically powered DD ... 40
Fig. 4.14 WDM-PON with optically powered DD ... 41
Fig. 4.15 TDM-PON with optically powered DD with WDM overlay for selective powering. ... 42
Fig. 4.16 TDM-PON with battery powered DD ... 43
Fig. 4.17 Output current and power vs. output voltage of arrays with 4 to 8 photodiodes illuminated in parallel and connected electrically in series ... 45
Fig. 4.18 Switching scheme in the initial startup procedure ... 47
Fig. 4.19 Block diagram of the optically powered DD ... 47
Fig. 4.20 Connecting an ONT to the network via a DD like shown in Fig. 4.19 ... 48
Fig. 4.21 Exploded view of a housing for a battery powered DD ... 49
Fig. 4.22 Setup drawing of the GPON testbed ... 49
Fig. 4.23 Inverted receiver output voltage showing measured response signals of the three polled DD in the GPON testbed shown in Fig. 4.22 ... 50
Fig. 4.24 Example network with base station and sensor nodes ... 51
Fig. 4.25 Optically powered sensor network and block diagram of a sensor node with opto-electronic power conversion ... 52
Fig. 4.26 Measured frame error ratio depending on the optical input power ... 53
Fig. 4.27 Switch consisting of a bipolar junction and a field effect transistor ... 55
Fig. 4.28 Start-up electronic for the optically powered sensor node. ... 57
Fig. 4.29 Current consumption of the whole start up electronic and of the voltage supervisor ... 57
Fig. 4.30 Leakage of an electrolyte capacitor measured as voltage and calculated leakage current ... 58
Fig. 4.31 Photograph of the optically powered sensor node and the attachable sensor submounts ... 60
Fig. 5.1 Schematic of data and power transmission over fiber ... 64
Fig. 5.2 (a) Unmounted single-cell PV converter (b) Enlarged view (c) Remotely available converted electrical power ... 65
Fig. 5.3 Optically powered remote unit with connecting fiber ... 66
Fig. 5.4 (a) Base station receiver diode connected to IO-ports of an FPGA-board (b) Eye diagram of decoded data after 200 m fiber length (c) Optically powered video transmission ... 67
Fig. 5.5 System overview ... 68
Fig. 5.6 Optically powered sensor platform ... 69

Fig. 5.7 Laser driver at the base station 70
Fig. 5.8 Optically powered sensor platform – receiver frontend 71
Fig. 5.9 Signal processing of the remote sensor platform 72
Fig. 5.10 SoC overview of the base station 73
Fig. A.1 Equivalent circuit of a real current source and an idealized photodiode. 82
Fig. A.2 Output current and power in dependence of the output voltage for the circuits shown in Fig. A.1. 83
Fig. A.3 Logic circuits 84
Fig. A.4 (a) Schmitt-Trigger (b) Connectors for the battery 84
Fig. A.5 Receiver consisting of photodiode, transimpedance amplifier and dual voltage amplifier 85
Fig. A.6 Transmitter with driver stage for (a) the DFB laser and (b) the VCSEL 85
Fig. A.7 Layout of demarcation device electronics with a DFB laser as transmitter in (a) top and (b) bottom view 86
Fig. A.8 Layout of demarcation device electronics with a VCSEL as transmitter in (a) top and (b) bottom view 86
Fig. A.9 Logic circuits 87
Fig. A.10 Receiver consisting of selectable photodiode, transimpedance amplifier, dual voltage amplifier with gain suppression, and comparator 88
Fig. A.11 Power supply consisting of photodiode array, voltage supervisor, switches for power supplies, and DC/DC boost converter 88
Fig. A.12 Layout of the sensor node electronics in top view 89
Fig. A.13 Layout of the sensor node electronics in bottom view 89

Zusammenfassung

Die vorliegende Arbeit befasst sich mit optisch versorgten Sensornetzwerken. In optisch versorgten Sensornetzwerken werden sowohl die Datensignale als auch die notwendige Versorgungsleistung der Sensorknoten über Glasfasern übertragen. Neben grundsätzlichen theoretischen Betrachtungen werden verschiedene Arten von Sensornetzwerken vorgestellt und untersucht. Die Art der jeweiligen Sensorapplikation entscheidet maßgeblich über Datenaufkommen, Komplexität und Energieverbrauch. Eine breite Spanne von Anwendungen wird demonstriert, um das Einsatzpotential solcher Netzwerke aufzuzeigen. Zusätzlich zum physischen Aufbau von Sensornetzwerken wird ein neues Kommunikationsprotokoll vorgestellt, das speziell auf optisch versorgte Sensornetzwerke angepasst ist.

Die alleinige Glasfaserverbindung zwischen der Basisstation und den Sensorknoten des Sensornetzwerks bringt einzigartige Vorteile mit sich: keine galvanische Kopplung, keine Störemission selbst bei hohen Datenraten, geringe Anfälligkeit gegenüber elektromagnetischer Störstrahlung und Einsatzmöglichkeit in Hochspannungseinrichtungen und explosionsgefährdeten Bereichen. Heutige Glasfasern haben eine geringe Dämpfung von nur 0,15 dB/km bei einer Wellenlänge $\lambda = 1{,}55\ \mu\text{m}$. Dies ermöglicht den Bau von weiträumigen Sensornetzwerken. Die Entwicklung von kostengünstigen Hochleistungslasern, effizienten photovoltaischen Wandlern und energiesparsamer und dennoch leistungsfähiger Elektronik während der letzten Jahre, haben die Weiterentwicklung von optisch versorgten Sensorknoten und -netzwerken gefördert.

In dieser Arbeit werden solche Sensornetzwerke näher untersucht. Geeignete Anwendungsbeispiele werden identifiziert und Prototypen für diverse optisch versorgte Sensorknoten werden vorgestellt. Dabei werden Knoten gezeigt, die sehr stark unterschiedliche Leistungsaufnahmen sowie Datenaufkommen besitzen. Des Weiteren werden die dazugehörigen Basisstationen vorgestellt. Diese Stationen müssen zum einen die Knoten mit Energie versorgen, zum anderen aber auch die Sensordaten entgegennehmen, anzeigen und speichern. Das eingesetzte Protokoll, das die Kommunikation zwischen Basisstation und Knoten regelt, ist auf die Topologie des Netzwerks abgestimmt. Es ermöglicht hohe Energieeinsparpotentiale, besonders bei Sensorknoten, die nur gelegentlich auftretende Messaufgaben übernehmen müssen.

Die photovoltaischen Wandler, die benötigt werden um die einfallende optische Leistung in elektrische Leistung zu wandeln, sowie deren elektronische Beschaltung wurden speziell für die unterschiedlichen Sensorapplikationen angepasst. Entscheidend bei der Auswahl der Komponenten und der Beschaltung war die mittlere Leistung, die im Wandler umgesetzt wird. Sensorknoten, die beispielsweise nur gelegentlich Temperatur- und Helligkeitswerte aufnehmen, können mit wenigen Mikrowatt mittlerer Leistung versorgt werden. Hingegen benötigen Knoten, die Videoinformationen aufnehmen und übertragen, mehr als 100 mW. Um die Ausgangsspannung des Wandlers an die benötigten Pegel für die elektronischen Komponenten anzupassen, werden Serienschaltungen von Wandlern untersucht und eingesetzt. Außerdem werden unterschiedliche Laser und Glasfasern für optisch versorgte Sensornetzwerke untersucht.

Unterschiedliche Implementierungen von Sensorknoten werden vorgestellt. En detail werden zwei stark unterschiedliche Knoten gezeigt.

Mit einer Leistungsaufnahme von 160 mW kann ein Knoten betrieben werden, dessen Videomodul 12,5 Bilder pro Sekunde in VGA-Auflösung aufnehmen und zur Basisstation schicken kann. Dazu wird der unkomprimierte Datenstrom von 160 Mbit/s einem direkt modulierbarem Laser aufgeprägt. Neben Sensorik umfasst dieser Knoten auch Aktorik. Zwei von der Basisstation aus gesteuerte Servomotoren können das Videomodul bewegen. Die Distanz zwischen Basisstation und Knoten beträgt 200 m und wird mit Multimode-Fasern überbrückt.

Im Gegensatz zum Videoknoten haben Sensorknoten, die nur gelegentlich Temperatur, Helligkeit oder Signalleistungspegel messen, eine extrem niedrige mittlere elektrische Leistungsaufnahme von 2 µW. Die Leistungsaufnahme kann stark reduziert werden, da der Knoten nur selten zeitliche kurze Mess- und Kommunikationsaufgaben übernehmen muss. Der Knoten wird daher die längste Zeit in unterschiedlichen Inaktiv-Modi betrieben, was zu einer äußerst geringen mittleren Leistungsaufnahme führt. Die Entfernung zwischen Basisstation und Knoten kann dabei mehr als 10 km betragen.

Basisstationen, welche die jeweiligen Sensorknoten mit Energie versorgen und deren Daten in Empfang nehmen, werden ebenfalls vorgestellt und diskutiert.

Beim Videoknoten werden die empfangenen Bilddaten von einem Field programmable gate array (FPGA) komprimiert und auf einer lokal betriebenen Webseite angezeigt. Ebenfalls kann auf der Webseite die Position des Videomoduls im Sensorknoten verändert werden. Dazu werden die entsprechenden Steuerinformation auf den Versorgungslaser aufmoduliert und im Knoten ausgelesen und ausgewertet.

Für die Sensorknoten mit geringer Leistungsaufnahme wird eine kompakte USB-versorgte Basisstation verwendet. Über den USB-Anschluss werden gleichzeitig die Sensordaten an einen Computer übermittelt.

Ein neues Kommunikationsprotokoll, das den Anforderungen eines optisch versorgten Sensornetzwerks entspricht, wird vorgestellt. In Netzwerken, die mehrere Sensorknoten beinhalten, werden passive Verzweiger eingesetzt, um die Knoten mit einer Punkt-zu-Multipunkt-Verbindung an die Basisstation anzuschließen. Diese faseroptischen Verzweiger verteilen Signale von der Basis an alle Knoten, Signale von den Knoten hingegen werden nur zur Basis übertragen. Energie-effiziente Carrier sense multiple access (CSMA) Protokolle, bekannt von drahtlosen Sensornetzwerken, eignen sich daher nicht für fasergebundene Netzwerke. Bei dem hier neu entwickelten Protokoll übernimmt die Basisstation die vollständige Kontrolle über das Kommunikationsmedium und teilt den Knoten Zeitschlitze zu. Knoten, die sich die meiste Zeit in Inaktiv-Modi befinden, werden durch spezielle Signale resynchronisiert und abgefragt. Resynchronisation ist notwendig, da im energie-effizientesten

Inaktiv-Modus der Knoten den Synchronismus mit der Basisstation aufgrund des dann unpräzisen Taktgebers verliert.

Als Anwendungsbeispiel wurden mehrere der Niedrigenergie-Knoten in ein faseroptisches Kommunikationsnetzwerk integriert. Die Knoten liefern dem Betreiber mehr Einzelheiten über das Netzwerk als mit klassischer Reflektrometrie möglich wäre. In diesen Netzwerken werden passive Verzweiger eingesetzt. Bei Messungen der reflektierten Leistung ergeben sich bezüglich der Position der Reflexionsstelle wegen der passiven Verzweiger mehrdeutige Resultate. Die Auflösung ist schlecht aufgrund der hohen Verzweigerdämpfung. Erfolgreiche Datenabfrage der Knoten bei zeitgleicher Datenkommunikation im Netzwerk wird erfolgreich gezeigt.

Unterschiedliche Basisstationen zusammen mit ihren dazugehörigen Sensorknoten bilden verschiedene Arten von optisch versorgten Netzwerken. Ergebnisse darüber wurden in verschiedenen Zeitschriften veröffentlicht [J1], [J3] – [J5] und bei internationalen Fachkonferenzen [C1] – [C9] vorgestellt.

Achievements of the Present Work

In this thesis, fiber-optically powered sensor networks have been investigated. Advantageous application fields have been identified and appropriate prototypes of different sensor nodes were developed, set up and characterized. Broad spans of power consumptions and data rates of the respective nodes were covered. Base stations to gather and display the information collected by the nodes and to supply the sensor nodes have been set up, too. Further, a special protocol taking into account passive optical fiber networks characteristics as well as different needs of different nodes has been developed. Improvements in power consumption of the nodes enabled through the special communication protocol were exploited.

In the following, we give a concise overview of the main achievements.

Studies on Optical Power Supplying: In general, sensor nodes have very different needs in power consumption and data throughput. Both strongly depend on the application. Here, simple temperature measuring nodes and complex video capturing nodes are presented. Average power consumptions span from few microwatts to several hundred milliwatt. For each node, the optical power supply was adapted to guarantee optimum performance. For this purpose, different photovoltaic power converters as well as lasers were selected and studied. Aside from the converter itself, the electrical circuiting of several converters and the appropriate impedance matching were studied to optimize the conversion efficiency especially for small optical powers. Different fiber types and different multiplexing techniques for power and data signals were implemented.

Sensor Node Design: Different sensor nodes were developed, set up and characterized. First, a group of nodes with an average power consumption of less than 2 μW is shown. Minimum power consumption is achieved by putting the nodes most of the time in an energy-saving sleep mode. A newly developed protocol allows addressability although sensors in sleep mode use inaccurate but energy-efficient clock sources.

Second, a node comprising a video module and servomotors for moving the module is presented. The power consumption of this node was measured to be 160 mW. The servomotors were controlled by the base station. Control signals were modulated on the high-power laser and detected in the node.

Base Station Design: Appropriate base stations for the above described nodes were set up. First, the base station for the low-power sensor nodes handled communication with the fiber-attached nodes autonomously and sent the gathered information via the network to a computer.

Second, a base station was developed for the node for capturing video data. The base station receives, processes and displays the video stream. The commands for controlling the servomotors in the node were also input there.

Studies and Developments in Protocols: Networks were composed of a base station and one or several sensor nodes. Passive power splitters were used to connect several nodes to one base station. These power splitters allow all nodes to listen to the base station. Signals from the node to the base station cannot be detected by other nodes. So, standard energy-efficient carrier sense multiple access (CSMA) protocols fail for organizing communication in such networks. Therefore, a newly developed protocol is presented combining random and scheduled access, so that the base station takes over full control of the communication. Occasionally polled sensors are re-synchronized with a special polling procedure which allows these nodes to stay in their most energy-efficient mode for the longest period of time. This operation mode leads to a loss of synchronism due to an inaccurate clock source used hereby. The protocol has been published in [J4].

Base station, nodes and protocol together were assembled in different network configuration to meet different application needs, namely:

Monitor Devices in Access Networks: One group of sensor nodes was integrated in a gigabit passive optical network (GPON) testbed for providing enhanced monitoring functionalities. Monitoring fiber failures with reflectometry measurements could suffer ambiguity and high losses due to high splitting ratios of the signals. These issues are solved with small versatile devices which operate independently from a power line at any point in the network. The monitoring procedure was run in-service and did not show a negative effect on the communication signals. [J1], [C3], [C4], [C7]

Sensor Devices in Fiber Networks: The nodes perform simple operations like measuring temperature, brightness and acceleration, and send the gathered information to a base station. Average power consumption of the nodes is reduced to less than 2 μW. The base station was powered by a standard USB port. It handled communication with the fiber-attached nodes autonomously and sent the gathered information to a computer. Power and data signals were transported over up to 10 km single mode fiber at wavelengths around 1550 nm. [J4], [C1], [C5], [C6]

Bidirectional Video System: This node comprises a video module and servomotors for moving the module. The 12.5 frames per second in VGA resolution are processed in a field programmable gate array (FPGA) and sent to a base station with a resulting data rate of 160 Mbit/s. The power consumption was measured to be 160 mW. This power was provided by a high-power laser at a wavelength of 808 nm. The servomotors were controlled by the base station. Control signals were modulated on the high-power laser and detected in the node. The generated data from the video module were transmitted to the base station. Another FPGA was deployed in the base station to receive, process and display the video stream on an interactive website. On the website also the commands for the servomotors were input. Base station and node were connected via 200 m of multimode fiber. [J3], [J5], [C2], [C8], [C9]

Summary

In optically powered networks, both communication signals and power for remotely located sensor nodes are transmitted over an optical fiber. Optical powering is a key enabler for a new generation of autonomous multifunctional intelligent subscriber and sensor networks with a broad range of monitoring and communication functions related to security of homes and public spaces, of roads, bridges and personal health as well as to general-purpose communications, to name just a few. One can also envisage optical powering of short-range passive optical networks (PON) comprising distributed link-supervision.

Key features of optically powered networks are node operation without local power supplies or batteries, operation with negligible susceptibility to electromagnetic noise and lightning due to galvanic isolation between nodes and base stations, operation in electrostatic discharge-sensitive environments, and operation without electromagnetic radiation from wires even at high and highest data rates. Last but not least it should be mentioned that optical fibers have very small attenuation, e. g., 0.15 dB / km for standard singlemode fibers. This opens the application field even for large network area coverage.

Despite the advantages of such networks, it is only in most recent years that advance towards inexpensive high-power lasers, highly efficient opto-electric converters and, most importantly, the advent of low-power high-performance electronics have alleviated the problem of limited local electric energy.

In this thesis, different kind of optically powered devices and networks are investigated and demonstrated with prototypes for thoroughly selected applications. The successfully implemented prototypes cover a broad span of power consumptions and data rates. Reduction in power consumption enabled by a newly developed protocol is exploited.

The thesis is structured as follows. In Chapter 1, the principle of optical powering is introduced and a short review is given. The fundamentals of communication protocols for networks with a shared medium are discussed.

In Chapter 2, the necessary theoretical background for opto-electronic power conversion is discussed. Physical and electrical aspects are addressed.

Chapter 3 comprises theoretical and measured results for a newly developed communication protocol. The protocol is tailored for the needs of an optically powered network. Aside from the special aspects of the fiber as a communication medium, heterogeneous sensor nodes of the network are taken into account. Heterogeneity results from different sensors attached at the respective node and manifests itself in transmitted data volume, priority, and latency. A combination of random and scheduled access allows handling of the heterogeneous nodes. Ratios between active and inactive periods down to 10^{-5} are demonstrated and result in an expected mean power consumption of a few microwatt. This research has been published in [J4].

One group of nodes was built with focus on minimal power consumption, and the results are discussed in Chapter 4. The mean electrical power consumption is reduced to less than

2 μW. Different types of sensors can be attached via digital and analog interfaces. The opto-electronic power conversion for supplying the node is optimized for powers in the microwatt range. The low-power nodes are proposed as new monitoring devices for an optical communication access network. A laboratory trial shows successful operation of the monitoring devices along with the communication network devices. Further applications of these monitoring devices in telecommunication networks are discussed with respect to the network topology. This research has been published in scientific journal papers [J1], [J3] – [J5], at international conferences [C1] – [C9] and in a patent [P2].

In Chapter 5, a sensor node with high power consumption and high data volume is shown. The attached sensor is a VGA camera module having a resolution of 640×480 pixels. With a power consumption of 160 mW, this node is able to capture 12.5 frames per second generating a data stream of 160 Mbit/s which is directly modulated on a laser and sent back to a base station. The base station hosts a true-time JPEG encoder and a webserver both realized on a field programmable gate array (FPGA). The captured video stream is displayed as Motion JPEG stream on an interactive website hosted on the webserver. Aside from the camera module, the node is equipped with servo drives to move the camera module. The drives are controlled via manual input at the interactive website.

Finally, an outlook on future research is provided.

1 Sensor Devices and Networks

This chapter first gives a general overview over different sensors that can be deployed, and how they differ in power consumption and in amount of data they deliver. Most important in a sensor network is the individual sensor node which will be discussed afterwards. Then, a brief look in former results of optically powered sensing will be taken. Networks consisting of several nodes and a base station will be introduced with focus on the possible heterogeneity of the nodes. Finally, existing protocols for communication and sensor networks will be reviewed. Especially, a closer look on energy-efficient protocols know from wireless sensor networks will be taken.

1.1 Fields of Application for Sensors

Environmental Sensors

Sensors measuring environmental parameters like temperature, brightness, humidity, acceleration, noise, voltage or current are called environmental sensors. These types of sensors normally have low power consumption (< 30 mW) and deliver data at low speed (< 100 kbit/s).

Still Picture Monitoring

The power consumption and the output data rate of image sensors is much larger than for environmental sensors. Image monitoring means the capturing and transmission of a few images per minute up to one image per second, so that only slowly varying processes can be monitored. Equipped with additional environmental sensors, a more complex sensor platform can be built up. For instance, an image could be taken if one of the environmental sensors detects unusual conditions. The average power consumption then can be as low as for environmental sensors if an energy buffer is integrated and the duty cycle for the image sensor is chosen properly.

Motion Picture Monitoring

In case the frame rate exceeds one image per second, the term video surveillance describes the application best. Power consumption and data rate increase significantly as the image sensor and the data logic need continuous powering. Power efficient systems are operated with 130 mW [38].

Actuators

Sensor nodes can be equipped with additional actuators. This opens another broad field of applications. In surveillance, the actuators can be used to redirect the image sensor's view, thereby enlarging the controlled area. Magnification and focus of the image sensor also can be changed. These actions can be controlled by the sensor node itself, by the base station or by the user at the base station. The latter cases need a communication link from base station to sensor node. In an optically powered system this can be easily realized by modulating the supply laser to encode data on it.

Power consumption of actuators often exceeds the power consumption of the sensing and the communication electronics. On the other hand, mechanical movement is not performed continuously at the sensor node, and the necessary energy might be buffered in accumulators or capacitors.

Radio Frequency (RF) Access Points

Different RF communication standards like WLAN [39], Bluetooth [40], ZigBee [41], GSM [42], UMTS [43] or LTE [44] can be transported over the optical link. The link also allows bridging large distances for power supply. The link ends in an optically powered node, where the signals are converted back into the RF domain and sent out through an attached antenna.

Robots

Finally, a combination of powerful logic circuits, different sensors and actuators gives the possibility to build an autonomously acting robot which is supplied from an optical cable and sends back its acquired information over the same cable. Such kind of robots can be used in areas where only a fiber connection would be feasible like in high-voltage installations, discharge-sensitive environments, and RF sensitive or RF polluted areas.

1.2 Optically Powered Devices

In Fig. 1.1 a schematic drawing of an optically powered sensor node and of a base station is shown. A laser located in a conventionally powered base station sends light through a fiber to the node. In the node, this light is converted back into electrical power which supplies (red arrows) electronics, sensors, and maybe even actuators. Data (blue arrows) gained from the sensors are transmitted back to the base station via an optical transmitter. The base station converts the optical data back into electrical signals. Both, power and data are converted from electronics (depicted with a green background) to photonics (yellow background) and vice versa. The connection between base station and node is just a glass fiber realizing a perfect galvanic isolation.

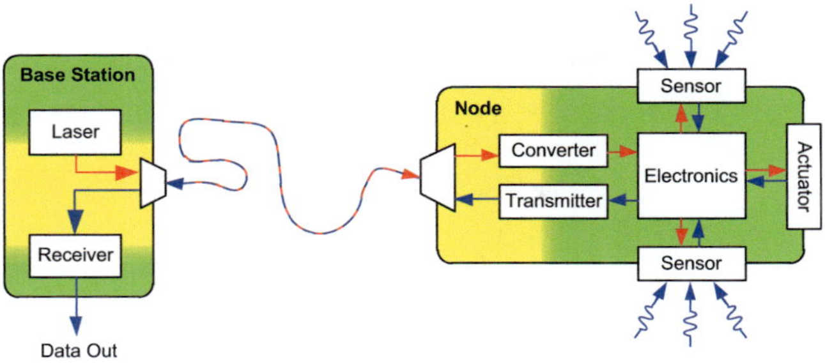

Fig. 1.1 Optically powered sensor node with base station. A laser located in an electrically powered base station sends light through a fiber to the node (red). In the node, this light is converted back into electrical power which supplies (also red) electronics, sensors, and actuators. Data (blue arrows) interrogated from the sensors are transmitted back to the base station via an optical transmitter, are received in the base station and converted back into electrical signals. Both, power and data are converted from electronics (depicted with a green background) to photonics (yellow background) and vice versa.

As optically powered devices operate without local power supplies or batteries, they can be deployed in discharge-sensitive environments. Furthermore, due to galvanic isolation such devices show negligible susceptibility to electromagnetic interference and lightning. Large distances can be bridged as a standard singlemode fiber has very small attenuation of only 0.15 dB/km. Although optically supplied, electrical power is available in the remote sensor node, and most of the common sensors can be operated. These sensors rely on an electrical power supply and return the sensor output as analog or digital electrical signal, or the desired quantity is transformed into an electrically measurable quantity. Thus, optically powered networks profit both from the unique advantages of a fiber as transport medium for energy and information, as well as from the variety of sensors available in electronics.

Review on Optically Powering [J4]

Historically, the basic idea of providing photonic energy to an optical network was realized as early as 1978 when an optically powered sound alerter was demonstrated at Bell Labs by DeLoach, Miller and Kaufman [4] for a fiber-based telephone network.One year later, the same laboratory reported satisfactory two-way speech transmission and vigorous sound alerting at the remote station with 14 mW of DC averaged laser power incident onto the fiber, Miller and Lawry [5], and this work was continued over the next three years [6]. A decade of development had passed when Kirkham and Johnston [7] from Jet Propulsion Laboratory provided in 1989 a brief historical review on fiber-optically powered devices along with the description of an optically powered data link with 1 kHz bandwidth for power system applications. It was ventured by Banwell et al. [8] from Bellcore in 1993 that it may be possible to operate conventional telephone station sets using electricity derived photovoltaically from light in a fiber. These authors concluded that with a power delivery to end-loop equipment below 30 mW for a 3.6 km loop (larger link attenuation) and 200 mW for a 1 km loop (smaller link attenuation) the cost of optical powering would be comparable to powering over copper. Further, an optically interrogated network of optically powered sensors was discussed by Pember, France and Jones [9] in 1995. They stored the photogenerated energy locally in a capacitor and emphasized the advantage of GaAs photogenerators (30 % efficiency) over silicon photodiodes (15 % efficiency). Another very specialized network consisting of optically powered electrical accelerometers was successfully built and field-tested by Feng [10] in 1998, while Pena et al. [11] described in 1999 an efficient power-delivery system for an all-fiber point-to-point connection with an optoelectronic sensor unit. With a viewpoint to overall energy saving, Miyakawa, Tanaka, and Kurokawa [12] published in 2004 design approaches to integrate solar energy harvesting in a power-over-optical local-area-network. The authors proposed a fiber optic power and signal transmission system for DC powering to such information tools as personal computers. Only recently, in 2007, the first optically powered video-link with bitrates above 100 Mbit/s was demonstrated [13].

Key to the success of optical powered networks is an energy efficient conversion of optical power into electrical energy. First serious studies on power budgets of optically powered links were performed by Liu [14] from DIAS (UK) in 1991 concluding that GaAlAs photovoltaic cells had to be chosen for a highly efficient power conversion. Record conversion efficiencies of 50.2 % were achieved in 2001 by van Riesen, Schubert, and Bett [15] from Fraunhofer Institute for Solar Energy Systems (Freiburg, Germany) when illuminating GaAs photovoltaic cells with an intensity of 6.5 W/cm^2 at a wavelength of 810 nm. The important influence of cell temperature on the conversion efficiency was emphasized by Miyakawa, Tanaka, and Kurokawa [16] in 2005. A most recent publication in 2008 by Werthen [17] discusses photovoltaic power converters (PPC) with electrical output powers over 1 W. The optimum light for a GaAs PPC lies in the wavelength range of 790 nm to 850 nm where various pump lasers with power levels as high as 5 W are available. For higher light levels above 10 W, InGaAs pump lasers in the range of 915 nm to 980 nm are most practical. To function with such lasers, the PPC also has to be made from a similar material, the bandgap of which and, hence, the voltage output from the PPC is less than that of GaAs devices. A PPC has been demonstrated with over 1 W of electrical output using a 5 W laser emitting at 960 nm. The authors remark correctly that at such output power levels even the remote powering of distributed antenna systems becomes possible.

Conversely, power consumption can be minimized at the sensor node. Lowest power device operation with microwatt-power InGaAs photogenerators for lightwave networks were pioneered in 1997 by Giles at al. [48] for powering a remotely-located optical shutter. For this application a 10 V optical-to-electrical InGaAs photogenerator was reported in 1999 by Dentai at al. [54] from Bell Labs.

1.3 Networks of Optically Powered Devices

A network of optically powered sensor nodes is shown in Fig. 1.2. All nodes are connected to the base station via fibers and power splitters. The resulting network has a tree-like topology. Signals coming from the base station (red) are split up and distributed to all nodes whereas signals coming from the sensor nodes are directed to the base station only. As a consequence, all nodes share one uplink channel but it is impossible for them to check whether another node is actually sending to the base station. This circumstance has to be considered when choosing an appropriate protocol for organizing communication between base station and sensor nodes.

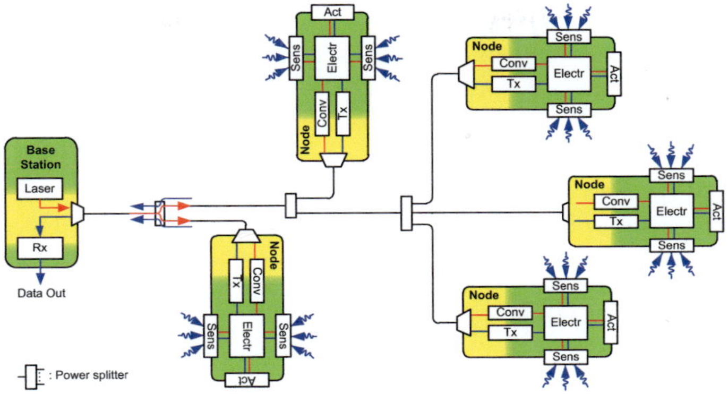

Fig. 1.2 Optically powered sensor network with its base station and nodes shown in Fig. 1.1. The nodes are connected to the base station with fibers and passive power splitters in a tree-like network. Signals coming from the base station (red) are distributed to all nodes by the splitters. Data signals (blue) coming from the node are combined in the splitters.

The exemplary network shown in Fig. 1.2 consists of identical sensor nodes which in reality is not necessarily true. A sensor network in general may consist of very different kind of nodes as has been discussed in Section 1.1. The nodes then differ in

- average power consumption,
- average daily data rate,
- duty cycle,
- priority,
- and latency.

The duty cycle τ is defined as the ratio between the length of an active time period T_{active} to the sum of active and inactive (T_{inactive}) periods:

$$\tau = \frac{T_{\text{active}}}{T_{\text{active}} + T_{\text{inactive}}} \tag{1.1.1}$$

Environmental sensors can be operated with duty cycles below 10^{-3}; motion picture monitoring requires a duty cycle in the order of 1. Different priorities may be implemented to distinguish between critical and non-critical nodes in the network. Latency is very important in real time applications like video-conferencing but is subordinate for most other data types.

These heterogeneous properties strongly influence the choice of the protocol. The following Section 1.3.1 discusses state-of-the-art media access control (MAC) protocols. These protocols control access to the shared communication medium, here access to the uplink channel. As the receiver and the transmitter consume most of the energy in a sensor node, their active times strongly influence the mean power consumption of a node. Switching off the communication interfaces has to be controlled by the MAC protocol to maintain connectivity.

1.3.1 Media Access Control (MAC) Protocols

A typical network consisting of several nodes and a central controller is depicted in Fig. 1.3. The nodes and the central controller share one transmission medium. If a node sends a message to another node or to the central controller, the medium is captured by this node. If a second node sends also a message which temporally overlaps with the first one, both messages interfere and cannot be received properly. A control mechanism is therefore necessary to avoid loss of data packets due to collisions of them in the shared medium. This is done in a MAC protocol. The following section discusses state-of-the-art for MAC protocols. An overview over the numerous approaches can be found in [26].

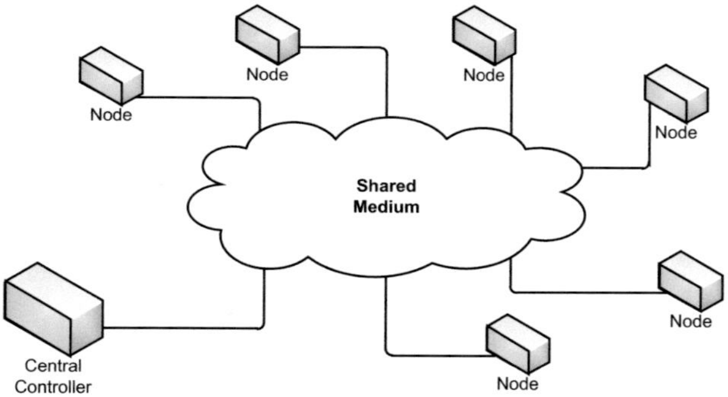

Fig. 1.3 Network consisting of a central controller and several nodes which share one transmission medium.

Different techniques of sharing the medium are categorized in Fig. 1.4. In the static channelization approach, the single nodes get their own communication channel within the medium, e. g., a separate wavelength in wavelength division multiplexed networks or a fixed time slot in time division multiplexed networks. However, nodes generating only little traffic will not make efficient use of their assigned channel. This makes the channelization scheme optimum for networks with constant and predictable traffic flows. Since in sensor networks the traffic is bursty and appears randomly, the channelization approach is unsuited here. Rather, a dynamic media access control is needed to use the given channel resources more efficiently. Two basic types for dynamic media access control are in use: scheduling and random access.

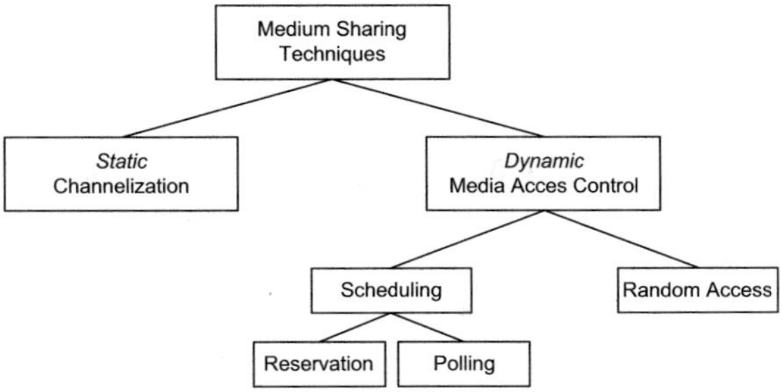

Fig. 1.4 Categories for medium sharing techniques. Modified from [26].

1.3.1.1 Scheduling

In scheduling approaches, access to the shared medium is organized in different ways. Here, two basic mechanisms of organization will be discussed: reservation and polling systems. In reservation systems, single nodes reserve a timeslot of fixed or variable length to capture the medium and transmit data. In contrast, polling systems have a fixed order. Node after node gets the ability to send. In both schemes, special data packets for organizing communication are necessary and have to be broadcasted to the nodes.

Reservation Systems

In reservation systems, traffic flow is organized in cycles. Each cycle begins with a reservation interval. Within this interval nodes can reserve a timeslot in which data will be transmitted. If timeslot lengths are chosen to be variable then also the length of the data packet has to be announced. After the reservation interval, the data transmission of the first registered node starts. The nodes are ordered in a fixed sequence, and send their data one by one according to their reservation.

With this scheme the medium can be used very efficiently in terms of data throughput. Only little overhead is needed to organize the communication, and nodes can quickly access the medium. However, in terms of energy efficiency this protocol is not very efficient as nodes are permanently powered up to be ready for receiving data or for requests of data.

Polling Systems

In contrast to reservation systems, the nodes take turns in accessing the shared medium in a polling system. After having finished, the active node passes the right to transmit over to the consecutive node in the queue. This passing is done by polling messages. Polling systems are often organized by a central controller which sends out the polling messages to the nodes. Nodes in turn signal the controller when their transmission has finished.

Again, only little overhead is necessary for organization, resulting in an effective data transmission protocol. But the nodes only know the sequence of the polling, and not the time when they are allowed to send. They have to stay in an active receive mode until a polling message directs a node to transmit data. In terms of energy efficiency, this is a major drawback.

Summary

Scheduling protocols allow quick access to a shared medium, and offer a high possible data throughput. However, the prize to be paid is permanently powering the devices. Sensor networks often have other requirements and needs in comparison to communication networks. Data throughput is not their major concern, because sensor nodes irregularly generate small amounts of data. A fixed schedule for communication is not needed. Here, random access protocols are more appropriate.

1.3.1.2 Random Access

The scheduling techniques described above use reservations or a fixed order to organize access to the shared medium in the network. The necessary messages for organizing traffic produce additional traffic within the network and lead to delays. Instead, nodes may access the shared medium whenever they have data to send. Due to its stochastic character this is called random access.

Nodes can send their data into the medium without sensing the carrier beforehand. A successful transmission to the receiver is then indicated by an acknowledge (ACK) message. Missing of ACK indicates a collision caused by at least one other node that sent a data packet at the same time. In this case the data packet is sent again. To avoid another collision with the second re-sending node, nodes do not re-send immediately after collision detection. They wait a random time until re-transmission (back-off algorithm). This very simple protocol was introduced already in the late 1960s and is known as ALOHA [26]. ALOHA does not try to avoid collisions of data packets, and under light network load, collisions are unlikely. With increasing load, collision probability strongly increases and protocols which avoid collisions are more suitable.

Collision probability can be reduced if the node senses the carrier before sending. If the carrier is found to be captured by another node, sending is shifted to a later time. Such techniques are called carrier sense multiple access (CSMA) protocols. Collisions cannot be fully avoided with this scheme. A node can sense the carrier as being idle, but an already sending node is not detected due to propagation delay of the signal. Collision detection and recovery also have to be included in these protocols. Before sending the whole data packet, the channel is captured with a defined sequence during a so-called contention period. If no collision is detected in the contention period, the actual data packet will be sent. Collisions will not affect long data packets, only the short contention messages might be lost due to collisions. This generates little overhead for these protocols, but it pays out especially under high load conditions where collisions are more likely.

All MAC protocols discussed up to now aimed at highest possible data throughput and easy accessibility to the network. Both are typical needs of a communication network. In contrast, sensor nodes in a sensor network irregularly generate small amounts of data and maximum throughput is therefore subordinate. Furthermore, energy consumption is not considered as a design parameter in standard communication protocols. However, energy is precious in optically powered networks and the whole network needs an energy-efficient design. The same holds true for wireless sensor networks (WSN). Here, the nodes are often supplied by a battery or an energy harvester. Both energy sources have limited capacities. So in this field, much research efforts are made to reduce power consumption of the nodes with appropriate protocols. Some of these approaches will be discussed in the following section.

MAC Protocols in Wireless Sensor Networks (WSN)

In WSN, random access protocols are widely used. They unite easy access to a shared medium and handling of little but bursty traffic. There is no need of energy-consuming upkeep of synchronism and schedules. This fits exactly to the demands of a WSN. Many different protocols have been developed and refined to meet these demands, and some of them will be briefly discussed in the present section. A focus will be set on protocols which mainly address energy-efficiency in terms of efficient use of the nodes' transceivers and logic circuits. An overview over the actual developments and trends in WSN MAC protocols can be found in [28].

Energy saving is done best by switching the node into a sleep mode without loosing synchronism. To communicate, the nodes can reactivate themselves synchronously. More reduction of power consumption can be achieved if synchronism can be dropped. In this case, another mechanism is needed to enable communication between nodes which are sleeping most of their time. Nodes can re-connect if one node waits in the active state until the other nodes also switch into the active state. The waiting node has to signal its need to transmit data by sending a long preamble. The other nodes wake up at regular intervals such that they become active at least once during the preamble, and wait until the data are sent, see Fig. 1.5 (a). This technique is called preamble sampling and was first introduced in 2002 [30], [31].

Still, a long preamble just signaling the other nodes that a message will be sent soon shows some disadvantages. With small duty cycles, the preamble is getting very long, and more energy is spent in the sending node for sending the preamble then for the data. The receiving nodes also waste energy with receiving the preamble signal constantly without listening to additional information. Furthermore, if only one node is addressed as recipient for to upcoming data transmission, all other nodes stay also in active mode until the end of the preamble. To overcome these issues, different variations of the preamble sampling techniques have been developed.

Improvements can be achieved by modifying the preamble: If the recipient node is addressed in the preamble all other nodes can switch back to sleep mode as there is no need for them to wait, see Fig. 1.5 (b), [32]. This avoids unnecessary and energy-costly receiving time for unaddressed nodes.

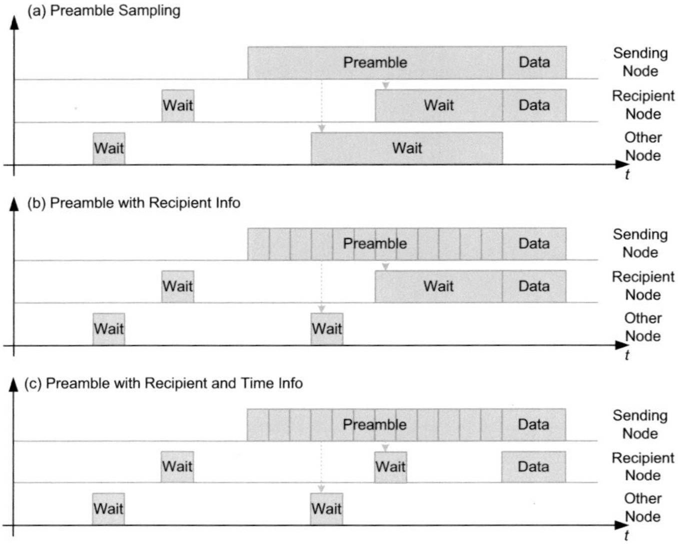

Fig. 1.5 Timing charts of different MAC protocols for WSNs (a) Preamble sampling: The sending node waits in active state until the other nodes also switch into active state to connect with them. The sending node signals its necessity to transmit data by sending a long preamble. The other nodes check for the preamble after having switched into active mode and wait until the data are sent. (b) Preamble sampling with recipient information: Like in (a), but additionally the recipient node is addressed in the preamble. Other nodes switch back to sleep mode as there is no need for them to wait. (c) Preamble sampling with recipient and receiving time information: Like in (b), but additionally the receiving time is also encoded to the preamble. The recipient node can bridge the time until data transmission in sleep mode. Drawing partially taken from [28].

The addressed node can save energy by additional encoding of the receiving time to the preamble. The recipient node can bridge the time until data transmission in energy-saving sleep mode, see Fig. 1.5 (c), [33]. This protocol is very energy-efficient for receiving nodes, but not for the sending node. If the sending node is interpreted as the base station of an optically powered network, we find a suitable communication protocol for this kind of network: the protocol is very energy-efficient for the nodes, power consumption is not critical in the base station, and the control of communication is found to be in the base station.

However, this protocol is not suited for networks of heterogeneous nodes. The medium is often blocked by preambles and high data throughput is not possible. Furthermore, only single nodes can be addressed in the preamble and broadcasting of data is not included. In Chapter 3, we present our low-energy MAC (LE-MAC) protocol which unites the capabilities of high data throughputs enabled by scheduling with the advantages of preamble sampling techniques.

Summary

When building a sensor network of optically powered sensors, following aspects have to be considered. The optical power needs to be converted efficiently into electrical power. Average power consumption for a sensor ranges from few microwatt to several hundred milliwatt. Appropriate components and designs are needed.

The necessary theory will be discussed in the following Chapter 2. In Chapter 3, a newly developed protocol for heterogeneous sensor networks will be presented.

2 Converting Light in Electrical Current

In this chapter, the fundamentals of opto-electronic power conversion will be reviewed. The operation principle of an illuminated photodiode will be discussed in terms of theoretical limitations and their cause. The dependence on the external load resistor connected to the photodiode will be examined, as well as possibilities to increase the output voltage by using a series connection of photodiodes.

The chapter is structured as follows. First, the theory of opto-electronic power conversion is discussed. Numerous textbooks focus on this topic, especially on the theory of sunlight conversion. A fundamental overview is given by Würfel [1]. More important for this work is the conversion of monochromatic light into electrical energy which is mainly discussed in scientific journals [2], [3], [64]. Here, opto-electronic power conversion in semiconductors will be discussed. A nice overview over the fundamentals of semiconductor theory may be found in the books of Sze [18], Ashcroft [19] and Kittel [20]. As a limiting quantity for conversion efficiency the saturation current of the absorbing p-n junction is identified and discussed.

Second, series connections of illuminated photodiodes will be introduced. They offer a higher output voltage as a single photodiode. Avoidable and unavoidable drawbacks of this method are identified.

Photon Absorption in a Photodiode

A photodiode is a semiconductor device with a p-n junction optimized to absorb photons. The current voltage characteristic of a p-n homojunction with current density J in the junction, saturation current density J_s, applied voltage U, ideality factor m and thermal voltage $U_T = kT/e$ with Boltzmann's constant k, elementary charge e and the junction's temperature T is

$$J = J_s\left(\exp\left(\frac{U}{mU_T}\right)-1\right) \qquad (2.1.1)$$

and results from minority carrier diffusion in the junction. The saturation current in the ideal diode has contributions from both electrons $J_{s,n}$ and holes $J_{s,p}$

$$J_s = J_{s,n} + J_{s,p} = en_i^2\left(\frac{D_n}{L_n n_A} + \frac{D_p}{L_p n_D}\right) \qquad (2.1.2)$$

with diffusion constants $D_{n,p}$ and diffusion lengths $L_{n,p}$ for electron and holes, and acceptor concentration n_A in the p-type semiconductor and donor concentration n_D in the n-type semiconductor. The intrinsic carrier density is n_i and strongly depends on the bandgap energy W_G of the semiconductor,

$$n_i^2 = N_C N_V \exp\left(-\frac{W_G}{kT}\right). \qquad (2.1.3)$$

Hereby, N_C and N_V are the effective density of states for the conduction and the valence band, respectively.

Between p-type and n-type semiconductor, an electric field is generated. If a photon with energy $> W_G$ is absorbed in the space charge region, one electron is moved from the conduction band to the valence band. A free electron hole pair is generated. As electron and hole have opposite sign of charge, the internal electric field accelerates the two carriers in opposite directions, so that immediate recombination is avoided. The field transports the electron into the n-type and the hole into the p-type semiconductor. In an outer connection between the two semiconductor materials (the attached load), a current can be measured.

Light coming with optical input power P_{opt} at an optical frequency f is assumed to be fully absorbed in the junction with light-sensitive area F. Each photon generates one electron hole pair causing one elementary charge e to flow through the outer connection. With these assumptions the optical input power and the generated photo current I_l are proportional,

$$\left(\text{Rate of photons} = \frac{P_{\text{opt}}}{hf}\right) = \left(\text{Rate of charges} = \frac{I_l}{e}\right), \tag{2.1.4}$$

where h is Planck's constant. The proportionality constant is called sensitivity S. In reality, only a part of the impinging photons generate electron hole pairs. The ratio is called quantum efficiency η and is assigned to the sensitivity

$$S = \frac{I_l}{P_{\text{opt}}} = \eta \frac{e}{hf}. \tag{2.1.5}$$

The generated photo current in the diode modifies the current voltage characteristics as additional current source in parallel to the diode. The equivalent circuit of an illuminated photodiode is shown in Fig. 2.1.

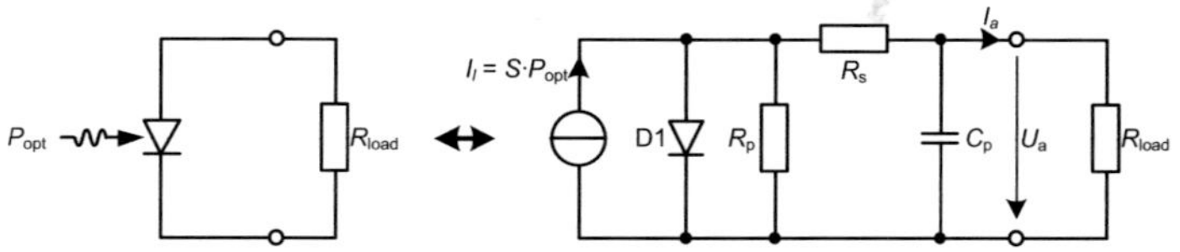

Fig. 2.1 Equivalent circuit of an illuminated photodiode. The generated photo current I_l is modeled as an ideal current source. The photon absorbing p-n junction acts as diode. Leakage current may flow in parallel to the diode, R_p. The series resistance R_s is due to bulk resistance. Parasitic capacities are lumped in the capacitor C_p. U_a is the output voltage of the photodiode seen by the attached load resistance R_{load}.

Leakage current may flow in parallel to the diode, indicated by a resistance R_p in parallel to the current source. The series resistance R_s is due to bulk resistance. Finally, parasitic and stray capacitors are lumped in the capacitance C_p. In the application field of this thesis the last three mentioned components of the equivalent circuit can be neglected for the following reasons: Leakage currents through R_p are normally neglected and the internal series resistance R_s is in the order of $1\,\Omega$ or even lower [22] and affects conversion efficiency only for high illumination intensities. As the photodiodes mainly are used as power converters and operated with direct current, capacitor C_p can be neglected, too.

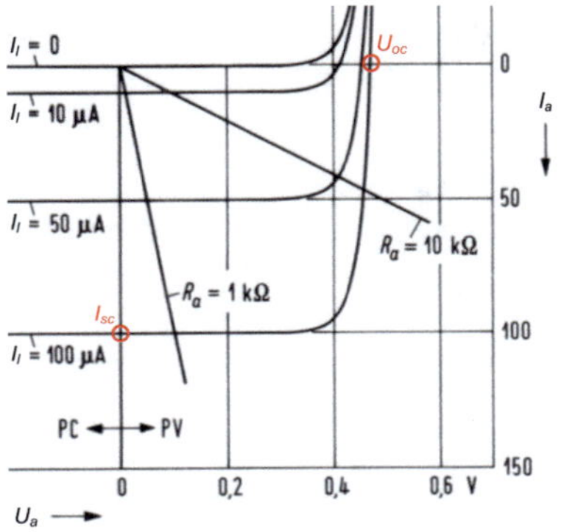

Fig. 2.2 Current voltage characteristic of a photodiode for different illumination powers. For one curve (I_l = 100 µA) short circuit current I_{sc} and open circuit voltage U_{oc} are marked with red circles. Taken from [81].

The output voltage U_a of the photodiode strongly depends on the load R_{load} attached to the photodiode. In Fig. 2.2, the output current I_a is shown in dependence of the output voltage. For 0 V all current generated by photon absorption flows as a short circuit current I_{sc} which equals the photo current

$$I_{sc} = I_l = SP_{opt}. \tag{2.1.6}$$

If the attached R_{load} is infinitely large, the photo current being a drift current must be compensated by an increased diffusion current. Therefore, the output voltage increases to the open-circuit voltage U_{oc}, which grows logarithmically with the incident power,

$$U_{oc} = mU_T \ln\left(\frac{SP_{opt}}{I_s} + 1\right), \tag{2.1.7}$$

where $I_s = FJ_s$ is the saturation current of the photodiode.

In Fig. 2.2, the linear (at $U_a = \text{const.}$) and logarithmic (at $I_a = 0$) dependence of the short circuit current and the open circuit voltage is illustrated. Operated at positive voltage and positive current, the output power $P_a = U_a I_a$ of the photodiode is positive and the photodiode generates electrical power. In this quadrant, operation of the photodiode is called photovoltaic (PV). Applying a negative voltage to the photodiode results in a negative power P_a, and the photodiode consumes energy. This operation is called photoconductive and is normally used for detecting optical signals. Here, we focus on the photovoltaic mode of the photodiode.

An exemplary current voltage characteristic is shown in Fig. 2.3 (a). In Fig. 2.3 (b) the corresponding output power is shown. For short and open circuit conditions, the output of the photodiode is zero. In between, the output power shows a maximum, the so-called maximum power point. To extract the maximum power from the photodiode, the attached impedance has to be matched. At the maximum, the differential of the output power equals zero,

$$\mathrm{d}P_a = \mathrm{d}(U_a I_a) = \mathrm{d}U_a \cdot I_{a,m} + U_{a,m} \cdot \mathrm{d}I_a \overset{!}{=} 0. \tag{2.1.8}$$

The matching condition (2.1.8) leads to

$$R_{\mathrm{load},m} = \frac{U_{a,m}}{I_{a,m}} = -\left.\frac{\mathrm{d}U_a}{\mathrm{d}I_a}\right|_m = -\frac{1}{G_D|_m}. \tag{2.1.9}$$

The load resistance $R_{\mathrm{load},m}$ where maximum power is taken from the photodiode equals then the differential resistance (= inverse of small signal conductance G_D) of the diode in the simplified equivalent circuit. This condition is comparable to impedance matching of a real current source where maximum power is taken when internal and attached resistance are equal. This is discussed in more detail in Appendix A.2.

Graphically, the load matching condition Eq. (2.1.9) lies on the curve where the tangent of the current voltage curve (= small signal conductance G_D) and a ohmic resistance curve $R_{\mathrm{load},m}$ (lines through origin) are perpendicular.

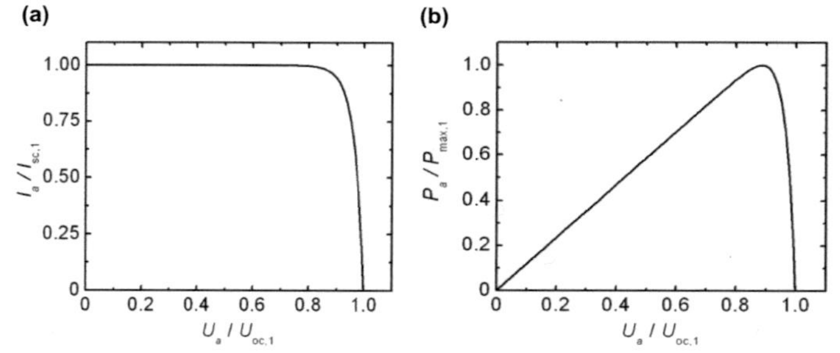

Fig. 2.3 Output characteristics of an illuminated photodiode (a) Normalized current and (b) normalized power density as function of normalized output voltage U_a. The diode is illuminated with an intensity of 10^5 W/cm^2.

Efficiency $\eta_{O/E}$ of the opto-electronic power conversion is defined as

$$\eta_{O/E} = \frac{P_a}{P_{\mathrm{opt}}}. \tag{2.1.10}$$

Aside from load matching, conversion efficiency can be improved by proper design of the photodiode. Efficiency is limited to

$$\eta_{O/E} < \frac{U_{\mathrm{oc}} I_{\mathrm{sc}}}{P_{\mathrm{opt}}} = S m U_T \ln\left(\frac{S P_{\mathrm{opt}}}{I_s} + 1\right). \tag{2.1.11}$$

Sensitivity S can be increased with a better quantum efficiency. However, improvement here is limited, as the quantum efficiency can not exceed 1.

Another possibility is decreasing the saturation current. Here, also a fundamental limit is given. This is discussed in more detail in Appendix A.1. Unavoidalble recombination as cause of the saturation current can be reduced with increasing bandgap energy W_G.

For monochromatic illumination of a photodiode, we want to give a theoretical upper limit of the conversion efficiency. Three assumptions are to be made. First, the saturation current is given by its theoretical lower bound defined by radiative recombination, see Eq. (A.18) in Appendix A.1,

$$I_{s,\mathrm{rad}} \approx e\frac{2\pi}{c^2h^3}F\exp\left(-\frac{W_G}{kT}\right)kTW_G^{\ 2}. \tag{2.1.12}$$

Second, quantum efficiency equals one, meaning absorption of all photons,

$$\eta = 1. \tag{2.1.13}$$

Third, the wavelength is chosen such, that the photon energy equals the bandgap energy to maximize sensitity,

$$S = \eta\frac{e}{hf} = \frac{e}{W_G}. \tag{2.1.14}$$

For these theoretical assumptions, the calculated efficiency over bandgap energy W_G and output voltage U_a is shown in Fig. 2.4. Efficiencies of up to 90% result. Devices optimized for illumination with laser light at a wavelength of 810 nm show measured efficiencies of more than 50% [3], [23]. Conversion efficiency is reduced by reflection, ohmic losses and non-radiative recombination.

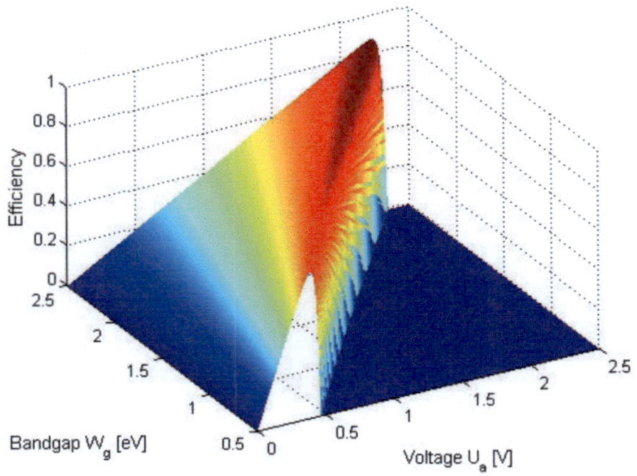

Fig. 2.4 Theoretical efficiency of an optimum photodiode in dependence of bandgap energy W_g and output voltage U_a of a photodiode for an intensity of 10^5 W/cm^2.

With increasing bandgap energy, also the maximum output voltage increases. An intermediate region is the near infrared region (0.78 – 3 µm according to ISO 20473, [24]) where lasers and photodiodes are available in a broad range of power and bandgap and where glass fibers show excellent performance in terms of attenuation. The corresponding bandgap energies in the near infrared are 1.60 – 0.42 eV. The maximum achievable voltage of 1.6 V is too small to run most of today's electronic devices, thus a voltage booster is necessary. One possibility is an electrical series connection of several illuminated photodiodes.

In Series Connected Photodiodes

Before the light is converted into electrical power it can be split, and the conversion can be done by several single photodiodes. If the photodiodes are connected electrically in series, their output voltages sum up, see Fig. 2.5. This approach was investigated with multi-segment photodiodes in 1979 by Borden [25].

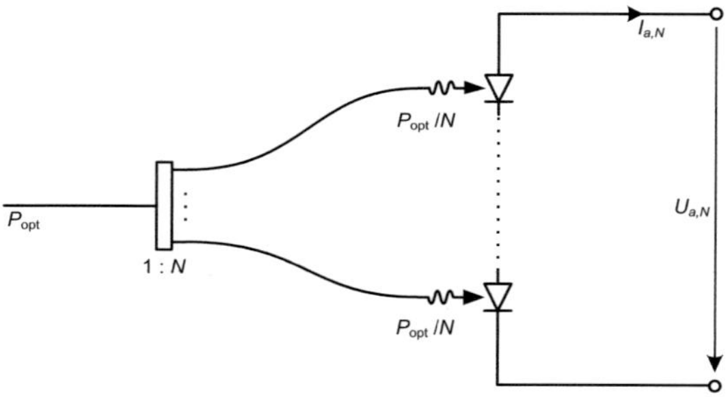

Fig. 2.5 The incoming optical power P_{opt} is equally distributed by means of a 1:N splitter to N photodiodes where the optical power is converted to electrical power. The photodiodes are connected in series to achieve a higher output voltage in comparison to a single photodiode illuminated with the full power.

We assume that the photodiodes are identical. As the incoming optical power P_{opt} is equally distributed among N photodiodes, the short circuit current $I_{sc,N}$ of the array is smaller by a factor of N in comparison to the short circuit current $I_{sc,1}$ of a single photodiode which is illuminated with the full power,

$$I_{sc,N}(P_{opt}) = S\frac{P_{opt}}{N} = \frac{I_{sc,1}(P_{opt})}{N}. \tag{2.1.15}$$

The open circuit voltages $U_{oc,1}$ of the single photodiodes add up and define the output voltage $U_{oc,N}$ of the photodiode array. The array has an open-circuit voltage of

$$U_{oc,N}(P_{opt}) = NmU_T \ln\left(\frac{SP_{opt}}{I_s N} + 1\right). \tag{2.1.16}$$

In the case of $SP_{opt} >> I_s N$, the term 1 in the natural logarithm can be neglected, and the open-circuit voltage of the array can be expressed using the open-circuit voltage of a single photodiode:

$$U_{oc,N}(P_{opt}) = NU_{oc,1}(P_{opt}) - mU_T N \ln N \tag{2.1.17}$$

The negative second term results from the fact that the open circuit voltage of one photodiode within the array is smaller than the open circuit voltage of a single photodiode receiving the same optical power as the whole array. Thus, $U_{oc,N}$ is smaller than $NU_{oc,1}$. The maximum available electrical output power of the array is therefore smaller than the one of a single photodiode,

$$U_{oc,N}I_{sc,N} = U_{oc,1}I_{sc,1} - mU_T I_{sc,N} \ln N. \tag{2.1.18}$$

In Fig. 2.6 the calculated current (a) and power (b) as function of the normalized output voltage for a single photodiode and arrays of 2, 4, and 8 photodiodes are shown. When increasing the number of photodiodes, the output current of the array decreases linearly. The output voltage increases, but less than linearly. This affects also the available output power: an array of photodiodes delivers less power than a single photodiode for same total optical input power. This reduction of efficiency is unavoidable when using an array.

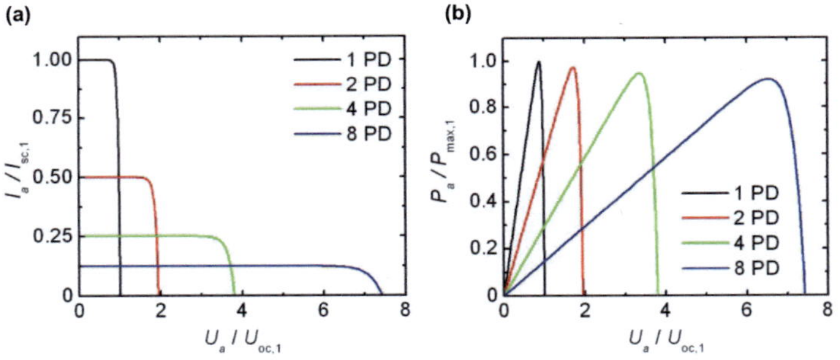

Fig. 2.6 Output characteristics of arrays of 1 (black), 2 (red), 4 (green) and 8 (blue) photodiodes (a) Normalized current and (b) normalized power as function of the normalized output voltage U. With increasing number of photodiodes the output current decreases and the voltage increases. The maximum electrical output power also decreases.

Furthermore, efficiency may decrease due to uneven illumination or uneven sensitivities of the photodiodes. In the series connection shown in Fig. 2.5, the generated photo current flows through all photodiodes. If for any reason, one of the photodiodes generates less current than the other photodiodes, the current through the whole series connection decreases to the smallest current in the series. This has to be considered when designing an array of photodiodes for power conversion with high output voltage.

3 Low-Energy Media Access Control (LE-MAC) Protocol

The results of this chapter were published in [J4]:

> ***Optically powered fiber networks***
>
> M. Röger, G. Böttger, M. Dreșchmann, C. Klamouris, M. Huebner, A. W. Bett, J. Becker, W. Freude, and J. Leuthold
>
> in *Focus Issue on Optics for Energy [invited], Optics Express*, vol. 16, is. 26, pp. 21821–21834, Dec. 2008. Reprinted, with permission, from [J4] © 2008 IEEE

In this chapter, we discuss an optically powered fiber network that connects and provides power to a multitude of subscribers, which are attached to a central office (CO) in a combined star and tree-like topology. Here, subscriber means the sensor node and CO the base station of a network. The focus of this chapter is on energy optimized subscriber hardware in combination with a new and flexible low-energy medium-access control (LE-MAC) protocol, which enables efficient use of the optically provided energy that is transmitted to each subscriber. Both, energy-hungry subscribers with high network priority (as is the need for video conferencing) and energy-preserving subscribers having a low network priority and a very small duty cycle (e. g., temperature or humidity sensors), can be handled by the CO simultaneously.

For an illustration, we refer first to a network of, e. g., temperature sensors. Temperature sensors need little power and typically are sampled only once in a while. Our LE-MAC protocol allows the sensors to accumulate and store energy within their idle time, then to perform a measurement for a short time, and to send the acquired information back. Multiple sensor modules are connected to one fiber. To avoid collisions of the sending sensors and to poll the multiple sensors on demand, the LE-MAC protocol organizes their idle and communication time slots.

Scenario of an Optically Powered Heterogeneous Network

For definiteness, we discuss exemplarily an optically powered subscriber network with representatives of the most important device types, see Fig. 3.1. Photonic power is distributed to a multitude of subscribers S_n $(n = 1, 2, \ldots N)$ with different power supply and bandwidth requirements.

The network consists of a line-powered intelligent central office CO (base station) with optical data transmitters (Tx) and data receivers (Rx) that are spatially or wavelength (de)multiplexed to (from) a single fiber. The CO transmitters supply data at a mean power level such that sufficient energy is transferred to the remotely connected devices. The subscribers feature an energy head comprising data transmitters and data receivers as well as a photonic-power receiver (Rp), all of which are spatially (de-)multiplexed to (from) the transmitting optical fiber.

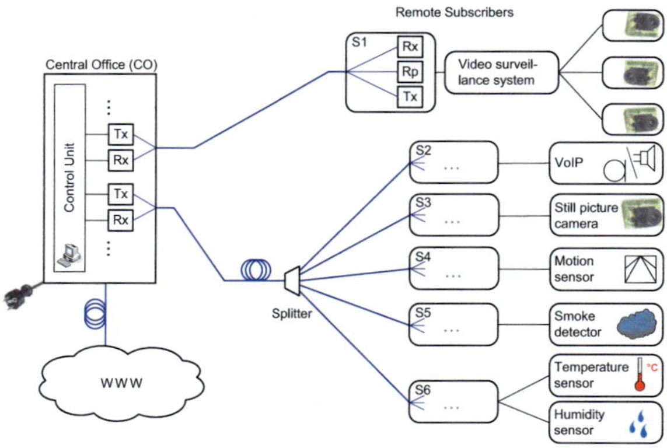

Fig. 3.1 Photonic network with optically powered subscribers. Optical transmitters (Tx) in the line-powered central office (CO, base station) transmit a downstream data signal with an appropriate average power for remotely supplying data and optical energy to subscribers. Optical receivers (Rx) in the CO sense the upstream data. Remote subscribers (S1…S6) comprise – besides the data acquisition and communication units – a section with optical data receiver Rx, photonic-power receiver Rp (photo-generator supplying electrical energy) and optical data transmitter Tx. A single Tx/Rx section in the CO can supply electrical energy to a multitude of subscriber sub-units. In the case of a video surveillance system (S1), sub-units would house various cameras at different locations or looking into different directions. Subscribers with high electrical power demand like S1 can be point-to-point connected to dedicated Tx/Rx units of the CO. Subscribers with lower energy demand are connected to the CO in a tree-like fashion and share one Tx/Rx of the CO. Such subscribers are for example voice over IP clients (S2), still picture cameras (S3), motion sensors (S4), smoke detectors (S5), temperature and humidity sensors (S6).

In this context, the designation "central office" comprises more or less complex base stations (e. g., "optical line terminations" (OLT)) that provide optical power along with data services, and "subscriber" stands for any remote device like sensors or general-purpose transceivers (e. g., "optical network units" (ONU), "optical network terminations" (ONT)), which are able to communicate with CO via the optical network.

Typical subscribers are compared in Table 3.1 with respect to their mean power consumption and their operating duty cycle. Here, duty cycle means the ratio of energy-costly active time periods, where measurement and communication tasks are performed, and idle time intervals spent in an energy saving (snooze) or even minimum-power (sleep) mode.

Table 3.1 Typical mean power consumption and duty cycle of different subscribers.

		Mean power consumption		
		low(few μW)	medium (few mW)	high (few 100 mW)
Duty cycle low	(10^{-5})	Temperature	Motion	
Duty cycle medium	(10^{-2})	Smoke	Still picture	
Duty cycle high	(10^{0})		Speech	Video

In the scenario of Fig. 3.1, low and medium power subscribers like speech communication (S2) using the voice-over-internet protocol (VoIP), still picture cameras (S3), motion detectors (S4), smoke detectors (S5), temperature and humidity sensors (S6) share a common fiber using remotely located active or passive power splitters. If need arises, subscribers with large mean power consumption like video conferencing or special surveillance systems (S1) can be supplied by the CO individually. The CO integrates all the heterogeneous subscribers in one network structure, and in addition provides an interface to the world-wide communication network (WWW).

The subscribers' heterogeneity and the specific network architecture – a combined star and tree-like topology – have important consequences for the communication between subscribers and CO: Subscriber signals can only be received by CO, and signals originating from CO must be broadcast to all subscribers. Therefore, a standard carrier sense protocol (for example, an energy-efficient version [34] of a carrier sense multiple access (CSMA) protocol) is not able to organize the communication. This is also true for the sensor-MAC (S-MAC) protocol [29] or for the low-duty cycle scheduled channel polling MAC (SCP-MAC) [35], both of which were designed for battery-operated wireless nodes. As a consequence, the CO's control unit alone has to organize the communication and all subscribers' needs regarding priority, bandwidth, and expected inactive times. In the following we describe a MAC protocol extension that meets the requirements of heterogeneous subscribers as envisaged in Fig. 3.1 and Table 3.1. With respect to low duty cycle subscribers, our protocol compares favorably with SCP-MAC insofar, as SCP-MAC has, for a given configuration, a minimum duty cycle for effectively reducing the power consumption ($3 \cdot 10^{-3}$ was demonstrated [35]), while our MAC protocol has not. We show experimentally that duty cycles as low as 10^{-5} are feasible, and that the lower limit for energy savings by lowering the duty cycle is given only by the devices' minimum energy consumption in sleep mode (for a duty cycle approaching zero).

Low-Energy Medium Access Control (LE-MAC) Protocol

In this section we present a low-energy medium access control (LE-MAC) protocol that serves the needs for optically powered heterogeneous subscribers in a simple and effective manner. To operate all subscribers with the least possible power consumption we extend specifications of common medium access control (MAC) protocols with the following features:

- Communication of subscribers only with CO
- Random and scheduled medium access of subscribers
- Quality of service with flexible assignment of priority and bandwidth
- Polling subscribers by CO broadcast, multicast and unicast replaces carrier sense
- Subscribers with high and low mean energy demand in one network
- Support of energy saving snooze mode: Subscribers maintain synchronism with CO.
- Support of minimum-energy sleep mode: Subscribers lose synchronism with CO. The sleep mode requires the following built-in features:
    - All communication circuitry switched off

- Autonomous wake-up needed, no external control possible
- Quick restoration of synchronism by listening to CO's polling at wake-up
- Reception of CO-scheduled rendezvous time
- Returning to snooze mode until wake-up at precise rendezvous time
- No energy-costly data transmission to CO before rendezvous time
- Communication with CO at rendezvous time
- Returning to sleep mode until next autonomous wake-up

The LE-MAC protocol's timing chart schematic is given in Fig. 3.2. The CO organizes the communication with subscribers by broadcasting the polling signals ① or ②, details of which are shown in the top row of Fig. 3.2.

The CO's communication protocol consists of alternating polling and Com sequences. A polling sequence comprises Fin, a Lstn and an Addr sequence, see ①. Optionally the polling signal might comprise a RV sequence, see ②. Details of the polling sequences are explained when discussing sequence ①.

In our example, the Fin sequence stops communication between any of the subscribers and CO. Then, subscribers S1 and S2 require CO's attention and send – after a random waiting time to minimize collisions – a request Rq to CO (broken arrows upwards). The CO then listens for a fixed time interval (Lstn) and acknowledges reception. At the end of the Lstn period the CO schedules the subscribers that have requested bandwidth. During the addressing interval (Addr, broken arrows downwards) the CO broadcasts addresses and timestamps for future unicast communication. Subsequently, the scheduled subscribers may communicate with the CO during the assigned communication time slot (Com, solid double arrow). Deferred subscribers learn by the broadcast when the next Fin or Lstn interval is to be expected, and can spend this idle time in an energy saving snooze mode before repeating their communication request Rq or listening to the next Addr information.

So far all subscribers maintain time synchronism with the CO, even in snooze mode. For lowest-power subscribers a precise quartz clock could be too energy-costly, so that subscribers with small duty cycle (S3 and S4 in Fig. 3.2) may reside in a minimum-energy sleep mode. Yet, while sleeping, only an inaccurate but ultra-low power clock is running for waking up the device, and so time synchronism with CO is lost. These devices cannot be scheduled accurately over a longer period.

A possible – but inefficient – communication with these subscribers could be as follows: A sleeping subscriber wakes up and either requests communication with the CO during the Lstn sequence, or checks the Addr sequence for scheduled communication. Since the wake-up time of the sleep mode subscriber is not accurate, the Addr request has to be repeated many times, and because sending data to the CO is energy-costly, this procedure increases the average power requirement of the subscriber. In addition, a considerable amount of bandwidth is wasted – particularly if there are many subscribers with sleep mode features.

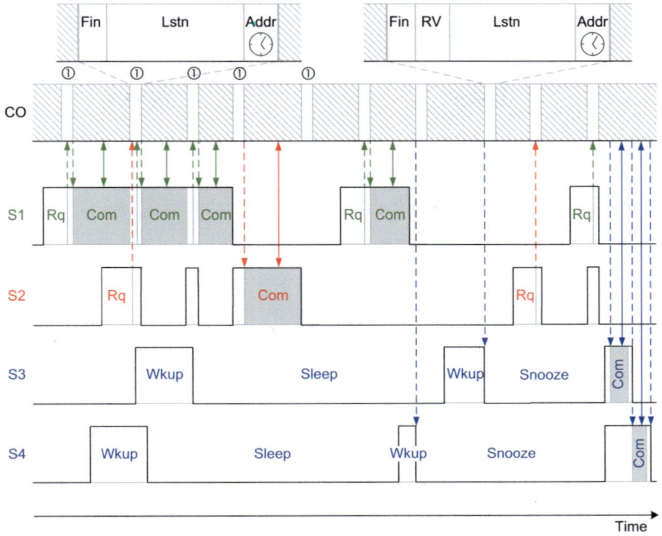

Fig. 3.2 Timing chart schematic for low-energy medium access control (LE-MAC) protocol; for details, see main text. Due to the treelike network architecture, the central office (CO) broadcasts its messages to each subscriber, which can communicate only with CO, but not peer-to-peer. – *Subscribers with high bandwidth demand* (S1, S2) along with low duty cycle subscribers (S3, S4) are handled by CO via broadcast polling (control) signals ① or ②, top row. Broken arrows stand for unidirectional, solid double-arrows for bidirectional communication (Com) of subscriber and CO. Access requests (Rq) of subscribers are queued by CO during the polling signal's listen interval Lstn in ①. Then CO decides which subscriber will be scheduled next and for what time interval. This is sent during the addressing interval Addr, after which communication can start (Com). – *Low duty cycle subscribers* spend most of the time in a minimum-energy Sleep mode that is interrupted by nearly periodically appearing wake-up intervals Wkup, which are initiated autonomously by the subscribers. Communication with CO is managed by broadcasting special polling (control) signals ②, top row right. During Wkup a rendezvous time stamp RV is sensed, a precise clock is set, and the subscribers return to a power-saving Snooze mode. At rendezvous time the subscribers awake, listen to be addressed, communicate with CO, and go again to Sleep mode.

Therefore, a more efficient protocol is needed. In order to save both energy and communication bandwidth, we introduce an additional rendezvous sequence (RV), the purpose of which is to efficiently inform sleep mode subscribers if and when a communication "rendezvous" will be arranged in not too far a future. The RV sequence typically would be a multicast call to a whole group of subscribers. Yet, it could be unicast as well as being a broadcast call. The protocol then would work as follows:

When waking up (WkUp in Fig. 3.2), subscribers S3 and S4 activate their basic receiver circuitry and listen for the CO's polling. If the rendezvous signal RV is *not* received during WkUp (polling signal ① in Fig. 3.2 as opposed to polling signal ②), the subscribers go back to sleep, as is the case for S3 and S4 during their first WkUp period in Fig. 3.2. However, if after the Fin sequence a device receives the rendezvous signal RV (polling signal ② Fig. 3.2), it extracts the time stamp for the next rendezvous with CO, sets a high precision clock to the rendezvous time, and then goes to an energy-saving snooze mode.

Snoozing subscribers (S3 and S4) maintain a precise quartz clock, awake exactly at rendezvous time and wait for being addressed by the CO. On reception of their individual address (broken arrows downwards), the first chosen subscriber S3 communicates with CO and exchanges data (solid double arrow). Having finished, S3 listens again to CO. On reception of a valid address other than its own (or being triggered by an internal time-out signal), S3 goes back to sleep mode. At this time (broken arrow downwards), S4 senses its own address, starts communicating with CO (solid double arrow), and ends the same way as formerly S3. This can be repeated for as many subscribers as needed. If a subscriber is not addressed or if the addressing signal is corrupted, an internal time-out mechanism sends the device back to sleep mode.

Beginning with the rendezvous time, communication requests Rq from higher-priority subscribers are deferred until the CO decides to end the interrogation of low duty cycle subscribers. It is also possible that – on command of CO – low duty cycle subscribers change their mode of operation and become attentive of polling signals ① in a manner described for the operation of S1 and S2, or that high priority devices fall back to low duty cycles and react to the rendezvous information RV in polling signals ②.

The allocation of bandwidth effected with polling signals ① and ② is very flexible. Subscribers with high priority (e. g., subscriber S1 in Fig. 3.2) can be preferred to subscribers with low priority (e. g., subscriber S2). Communication with low duty cycle devices (subscribers S3 and S4) can be also arranged at the discretion of CO. Thus a low-latency priority-driven quality of service feature is integral part of the LE-MAC protocol.

The most challenging part of the LE-MAC protocol is communication with low duty cycle subscribers that spend most of their time in sleep mode. During this time, the devices cannot be addressed by CO and lose time synchronism as described earlier. To validate the design of this part of our protocol we set up a network of four ultralow duty cycle subscribers S3 … S6 and a CO. For avoiding unnecessary complications we connected CO and subscribers by a wired network, the topology of which is shown in Fig. 3.3.

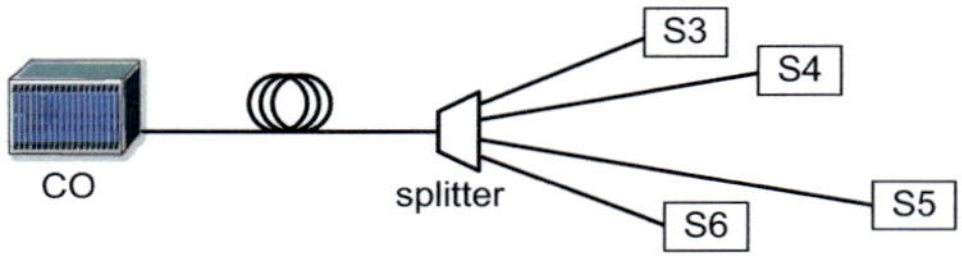

Fig. 3.3 Network of ultralow duty cycle subscribers S3 … S6 connected to a central office CO

The experimental CO and the subscribers were realized each with a mixed signal microcontroller μC from the Texas Instruments MSP430-family and powered with a supply voltage $V_b = 3.6\,\text{V}$ [36]. These μC are designed for sensor systems that capture analogue signals, convert them to digital values, and then process the data for transmission to a host system. One microcontroller serves as CO, is therefore always kept active and maintains an accurate clock. The other μCs act as subscribers. The devices communicate by exchanging serial data via their inbuilt universal asynchronous receiver / transmitter (UART) units [37].

Four subscriber operating modes were experimentally tried, and the total supply currents I_b are listed in Table 3.2. In *sleep mode* the μC maintains an internal low-quality (low-Q) very low power clock ($I_{Sleep} = 0.5\,\mu A$), while in *snooze mode* an external high-quality (high-Q) 32 768 Hz quartz crystal clock ($I_{Snooze} = 1\,\mu A$) is active. In *active mode* the μC operates with a digitally-controlled oscillator (DCO) frequency of 8 MHz ($I_{\mu C} = 3.5\,mA$). The internal DCO provides a fast turn-on clock source and stabilizes in 1 μs, however, the external quartz clock needs about 60 ms settling time. I_{Sleep}

Table 3.2 Subscriber operating modes and total DC supply currents I_b for a supply voltage $V_b = 3.6$ V

Subscriber modes		Attributes	μC current	Supply current	Total I_b
Sleep	μC low-power modes	Low-Q clock	0.5 μA	I_{Sleep}	0.5 μA
Snooze		High-Q clock	1 μA	I_{Snooze}	1 μA
Receive	μC active mode	Receiver on	3.5 mA	I_{Rx}	6 mA
Transceive		Transceiver on		I_{RxTx}	11 mA

In addition, the *receiving mode* and/or *transmitting mode* require external (optical) receiver (Rx in Fig. 3.1) and/or (optical) transmitter circuitry (Tx in Fig. 3.1) to be switched on, and this increases the supply current – depending on the actual Rx and Tx design – to practical values of $I_{Rx} = 6\,mA$ (receiving) and $I_{RxTx} = 11\,mA$ (transceiving), respectively. The *transceiving mode* comprises both an activated receiver and transmitter.

In Fig. 3.4 we show a measured timing diagram for our experimental network Fig. 3.3 which consists of a CO and four ultralow duty cycle subscribers S3 … S6. The upper curve labeled CO shows the transmitter signal of the central office. For our experiments, the control signal ② of Fig. 3.2 has been simplified and reduced to its rendezvous (RV) part, which has the duration T_{RV}. Inside a period that must be smaller than the subscribers' smallest sleeping time T_{Sleep}, the CO broadcasts to the subscribers R rendezvous signals repeated at intervals $T_R = T_{Sleep} / R$. The RV signal (details see left inset of Fig. 3.4) comprises a time stamp which tells each listening subscriber the waiting time T_{Snooze} that can be spent snoozing until the next rendezvous with CO.

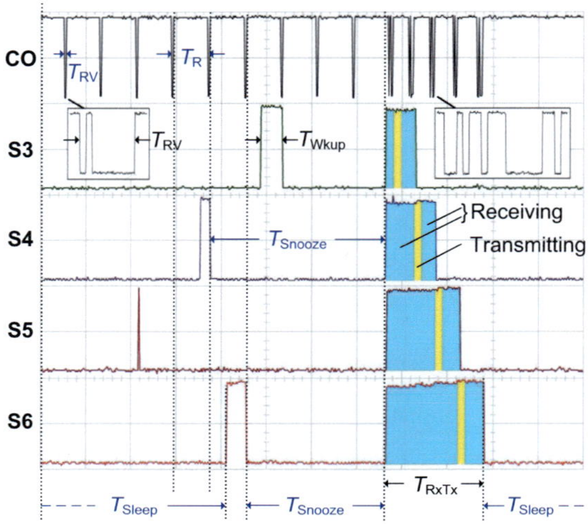

Fig. 3.4 Measured timing chart for low-energy medium access control (LE-MAC) protocol. Four randomly self-activating subscribers (S3…6) synchronize and communicate with the central office CO. Trace CO displays the data transmitter voltage of the CO. High S3…6 levels indicate energy-costly receive or transceive (RxTx) modes, low levels mark low-power Sleep or Snooze modes. Within a time interval T_{Sleep} the CO polls the subscribers with rendezvous signals RV (width T_{RV}) that repeat with a period $T_{\text{R}} = T_{\text{Sleep}} / R$ where R is a fixed number. These RV signals transmit information about the next rendezvous time with CO. If a sleeping subscriber becomes awake during a rendezvous signal is broadcast by CO, the subscriber senses the rendezvous time, starts a precise clock and goes snoozing. At rendezvous time all subscribers awake, wait for being addressed and exchange data with CO. An ending signal issued by CO sends the subscribers back sleeping.

Each subscriber stays sleeping as long as possible for the envisaged application, and its CPU remains disabled for about the sleeping time T_{Sleep}. Then an internal low-Q clock interrupt awakes the CPU, which, having activated the external high-Q quartz clock, waits for 60 ms until the clock has stabilized. Next the receive mode is activated for a time $T_{\text{Wkup}} > T_{\text{R}}$, and the subscriber stays listening whether CO broadcasts a rendezvous signal RV, see Fig. 3.4 trace S3. On reception of RV, the subscriber switches to an energy-saving snooze mode, watches its high-Q clock and activates itself precisely at the scheduled rendezvous time. If no valid RV signal could be decoded, the subscriber goes back sleeping.

At rendezvous time all subscribers switch to receive mode and stay listening for their address to be broadcast by CO (signal details see right inset of Fig. 3.4). When for instance subscriber S4 recognizes its address (transition from the dark (blue) to the light (yellow) region in Fig. 3.4), it activates its transmitter, sends to CO whatever information was required, and deactivates its transmitter again (transition from the light (yellow) to the dark (blue) region in Fig. 3.4). If CO fails to receive the subscriber's data, CO repeats addressing S4 a fixed number of times until either reception succeeds, or until the subscriber's time-out mechanism sends S4 back sleeping. When CO received the subscriber's data (or gave up interrogating S4), CO broadcasts the next subscriber's address (S5 in our case), and the previously addressed subscriber S4 returns to sleep mode as soon as it recognizes a valid address not being its own. The data exchange process repeats as described for the time interval T_{RxTx} that depends on the discretion of CO. Having interrogated all desired subscribers, CO broadcasts an End command (not marked in Fig. 3.4) thus sending the remaining listening subscribers to sleep mode.

For an estimate of the subscriber's energy consumption we determine the average supply currents I_{b} in the various modes assuming the following parameters, see Fig. 3.4: CO polls the subscribers periodically in intervals $T_{\text{Poll}} = 30\,\text{min}$. All subscribers wake up randomly inside a time interval with length $T_{\text{Sleep}} \approx 600\,\text{s}$. When polling, the CO broadcasts $R = 20000$ rendezvous signals with a period of $T_{\text{R}} = T_{\text{Sleep}} / R = 30\,\text{ms}$. Consequently, the subscribers need staying in receive mode for an average wake-up time of $T_{\text{Wkup av}} = T_{\text{R}} / 2 = 15\,\text{ms}$. At rendezvous time, the longest data exchange lasts $T_{\text{RxTx}} = 5\,\text{ms}$. With these assumptions, the various duty cycles for wake-up, data exchange and polling times are

$$\tau_{\text{Wkup}} = \frac{T_{\text{Wkup av}}}{T_{\text{Sleep}}} = 2.5 \cdot 10^{-5},\; \tau_{\text{RxTx}} = \frac{T_{\text{TxRx}}}{T_{\text{Sleep}}} = 8.3 \cdot 10^{-6},\; \tau_{\text{Poll}} = \frac{T_{\text{Sleep}}}{T_{\text{Poll}}} = 0.33 \qquad (3.1.1)$$

With the data provided by Table 3.2, the average supply current I_b may then be estimated,

$$\begin{aligned} I_b &= (1 - \tau_{\text{Poll}})(I_{\text{Sleep}} + 2\tau_{\text{Wkup}} I_{\text{Rx}}) \\ &\quad + \tau_{\text{Poll}}(\tfrac{1}{2} I_{\text{Sleep}} + \tfrac{1}{2} I_{\text{Snooze}} + \tau_{\text{Wkup}} I_{\text{Rx}} + \tau_{\text{RxTx}} I_{\text{RxTx}}). \end{aligned} \qquad (3.1.2)$$

There is no local minimum for the average supply current I_b as opposed to [35], only a lower bound $I_{\text{b low}} = I_{\text{Sleep}} = 0.5\,\mu\text{A}$ if the duty cycles approach zero, $\tau_{\text{Wkup}}, \tau_{\text{RxTx}}, \tau_{\text{Poll}} \to 0$ for $T_{\text{Wkup av}}, T_{\text{TxRx}} >> T_{\text{Sleep}} >> T_{\text{Poll}}$. For the realistic operating parameters chosen in Eq. (3.1.2), all subscriber modes as listed in Table 3.2 contribute about equally (some $10^{-7}\,\text{A}$) to the total average supply current. It amounts to $I_{\text{b}} = 0.86\,\mu\text{A}$, hardly more than its lower bound $I_{\text{b low}}$. With a supply voltage of $V_{\text{b}} = 3.6\,\text{V}$ the average electrical power per subscriber is 3 μW.

Our low-energy medium access control protocol is designed specifically for optically powered devices. For demonstration purposes some of the above described functionalities have been chosen. The following Chapter 4 shows results of a network of optically powered senor nodes and a possible application scenario in a fiber communication network. Different sensors, from simple temperature measurements to ambitious video streaming, combined with actuators like servo drives are discussed in Chapter 5.

4 Applications with Low Power and Low Bandwidth Demand

The following section discusses results of optically powered sensor networks consisting of different low-power sensor nodes. A special application scenario like monitoring of an optical access network will be discussed in Section 4.1, as well as general issues in Section 4.2. These results have been published in [C4], [J1],[C1], [S2], [S6] – [S8].

4.1 Advanced Monitor Device in Access Networks

The results of this section were partially published in [C4] and studied in a Bachelor Thesis and in a Project Report [S2], [S6].

The prevailing access network technology will be based on extended-reach passive optical networks (PON) combined with wavelength division multiplexing (WDM) techniques. Such networks connect cabinets, buildings or even homes to a central office (CO) and are denoted as fiber-to-the-*x* systems (FTT*x* with *x* = C, B, H for *c*abinet, *b*uilding, *h*ome). They are expected to reduce cost in the optical access range, and to provide significantly increased data rates. For the operators and subscribers, network availability as well as network security will become increasingly important, and operators ask for solutions to monitor and protect their infrastructure on the physical layer [73] – [76].

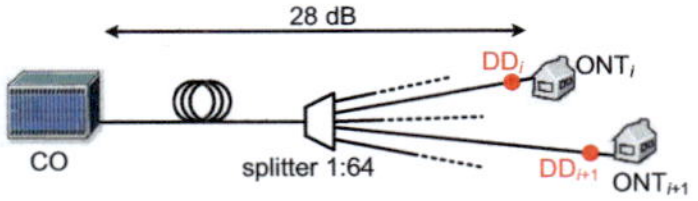

Fig. 4.1 Demarcation devices DD*i* (•) in an FTT*x* network for monitoring the operator-subscriber interface to ONT*i*. Figure reprinted from [C4] © 2010 OSA.

At so-called demarcation points different optical network segments have to be separated to define the responsibilities of operator and subscriber. For a PON system operator, there is presently no means available to assure proper performance of a fiber link up to the customer premises unless an active optical network termination (ONT) is attached to the line. Therefore, we propose using a traffic-transparent energy-autarkic demarcation device (DD), which is located at the demarcation point (see Fig. 4.1) and can be interrogated by the CO [73], [77].

An energy-autarkic DD could draw its energy either from a long-life (20 years) battery, or from optically transmitted energy transmitted via one dedicated wavelength channel to a large number of DD [72]. In both cases the electrical power requirements need be minimum.

Here, we describe a low-energy DD supplied with a 5 μW optical power channel or a long-life battery, and an appropriate protocol. The DD is transparent for data traffic, but responds to a low-bitrate control signal which the CO superimposes to the optical data stream.

The demarcation device connects the CO via the operator's network with the subscribers' premises, see Fig. 4.2. The CO sends pay-load data (3 dBm) to the subscriber (–26 dBm) at a wavelength λ_{data}. In addition, this signal carries a low-bitrate intensity modulation (modulation depth 10 %, ≤ 1 Mbit/s, UART protocol), by which the CO transmits control data to the DD. The perturbation caused by the control data is small enough to avoid frame loss at the ONT as was confirmed by true-traffic measurements. A 5 %-tap connects the data stream to a low-noise receiver (Rx, –37 dBm), which decodes the control data. A directly modulated laser transmitter (Tx, –3 dBm, λ_{DD}) sends information back to the CO via another 5 %-tap. All units are operated by a low-power microcontroller (µC).

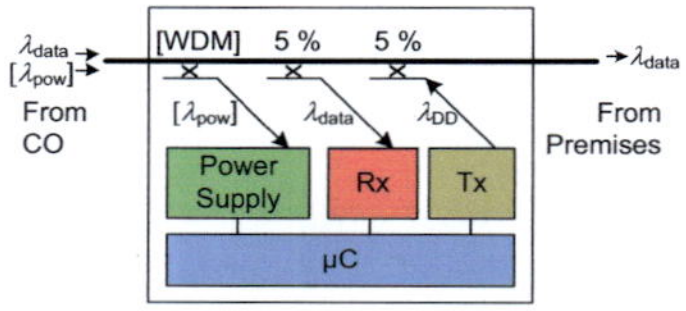

Fig. 4.2 Block diagram of the DD. The DD is placed before the customer's premises and taps out 5 % of the data signal to receive (Rx) control signals from the CO and send (Tx) status information back. A microcontroller (µC) is responsible for power management and data communication. Power can be delivered by an optical supply channel on a dedicated wavelength λ_{pow} which will be fully coupled out by a WDM coupler or by a long-life battery. Figure reprinted from [C4] © 2010 OSA.

Receiver Circuit

Being addressable is necessary for the DD to exchange data with the CO. Therefore a receiver circuit was designed to detect the control signal. This signal is encoded as small amplitude perturbation on the downstream signal going from the CO to the customer. A modulation depth of 5 – 10 % does not yet cause additional frame loss in the downstream signal [78]. The resulting total input power for the receiver ranges between 80 nW and 2.5 µW and signal photo currents will be between 4 and 125 nA assuming the sensitivity of the photodiode to be 1 mA/mW. The photo currents have to be converted and amplified to voltage signals and fed to the microcontroller. The microcontroller needs voltage signals with several 100 mV voltage swing to extract the binary encoded data. The bandwidth of the receiver needs to be at least 1 MHz and the current consumption should not exceed 10 mA to meet hard efficiency constraints.

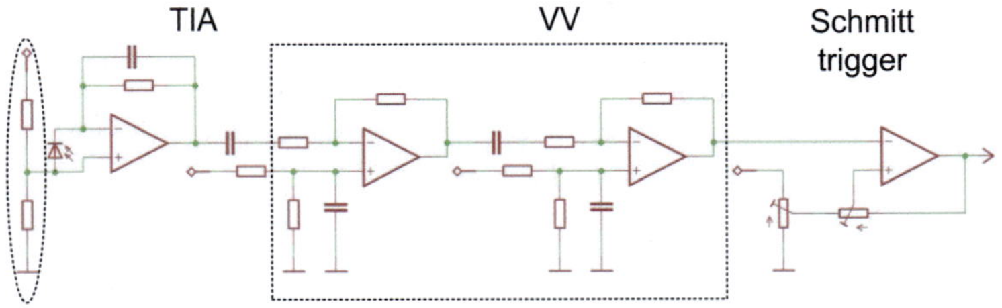

Fig. 4.3 Receiver circuit consisting of transimpedance amplifier (TIA), inverting voltage amplifiers (VV), and Schmitt trigger circuit. Decision thresholds of the Schmitt trigger can be chosen with two potentiometers.

A circuit fulfilling all above mentioned needs is shown in Fig. 4.3. It consists of a transimpedance amplifier (TIA), to inverting voltage amplifiers (VV) and a Schmitt trigger circuit. The voltage divider in front of the TIA shifts the operation point of the TIA to 1 V. This is necessary as only an asymmetric power supply for the operational amplifiers is available. Additionally, the voltage divider increases the transimpedance to a total of $498\,\mathrm{k}\Omega$. The VV show together an amplification of 100. The Schmitt trigger adapts the incoming signal to the voltage levels of the microcontroller and guarantees sharp edges.

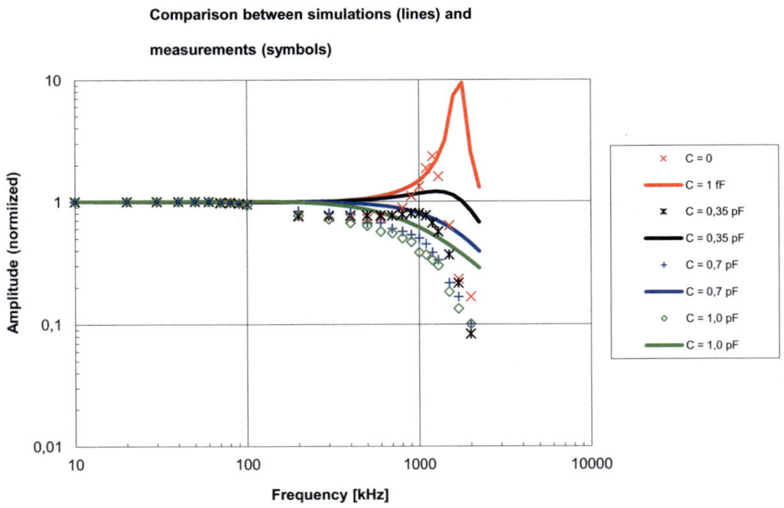

Fig. 4.4 Transfer function of the TIA, measured and simulated with different capacitances in the feedback of the TIA. Measurement (solid lines) and simulations (symbols) show good agreement within tolerances.

The capacitor in the feedback loop of the TIA in Fig. 4.3 is necessary for stable operation of this amplifier stage. However, this capacitor limits the bandwidth of the amplifier and has therefore to be chosen as small as possible. Simulations performed with Cadence® Pspice®

[79] and measurements for different capacitances are shown in Fig. 4.4 and show good agreement within tolerances and the very small values of the used capacitors.

Measurement results for the whole circuit are shown in Fig. 4.5. The transfer function of the receiver circuit is depicted in (a). The photodiode was illuminated with 80 nW (-41 dBm) optical power with a sinusoidal modulated with a depth of 10 % to the downstream signal at 1.24 Gbit/s and a pseudo random bit sequence (PRBS) of length of 2^7-1. Upper cutoff frequency is determined to 1.7 MHz. This cutoff results from the deployed low-power operational amplifiers and the chosen amplification. The lower cutoff frequency of 25 kHz is due to the AC coupling of the different amplification stages. The chosen capacitance of the coupling capacitors is a compromise between short switch-on delay and small lower cutoff frequency. The noise floor is depicted in grey. Main contributor to the noise is the TIA as first amplifier in a cascade of amplifiers [80]. Fig. 4.5 (b) shows the output signal of the VV in the case that the sinusoidal is replaced by a square-wave signal with a frequency of 500 kHz. This corresponds to a 1010… bit sequence at the target bit rate of 1 Mbit/s.

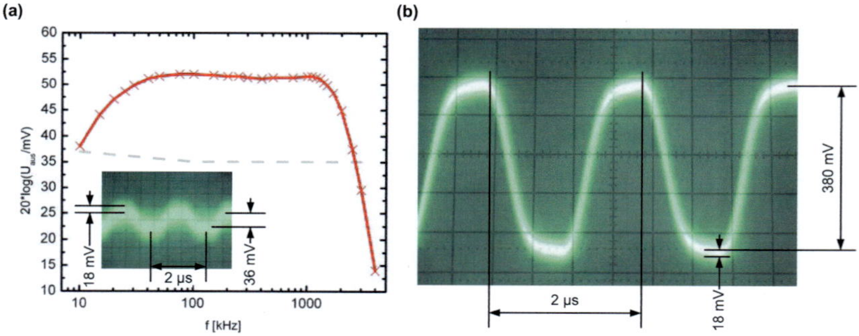

Fig. 4.5 Measurements on the performance of the receiver circuit at 80 nW total input power (a) Transfer function (red, solid) and noise level (grey, dashed). The inset shows the output voltage of VV at an optical input power of 5 nW. The input signal was a square-wave signal with 500 kHz with 10 % modulation depth. (b) Same as inset but with total optical power of 80 nW.

The receiver performance also depends on the downstream signal from CO to customers. The receiver works error-free within the specified dynamic range if the downstream is chosen to 1.24 Gbit/s with a PRBS of length 2^7-1. In Fig. 4.6 the effect of a longer, $2^{31}-1$, PRBS can be seen. The noise amplitude has increased significantly. Fig. 4.6 (a) shows the voltage signal at an optical power of 8 nW to the photodiode, again 10 % modulation depth with a 500 kHz square-wave signal. Sensitivity limit is reached as the signal to noise ratio (SNR) equals 1. Fig. 4.6 (b) is taken at an optical input power of 80 nW with the same modulation. A SNR of 10.7 results, corresponding to a bit error probability of $5 \cdot 10^{-4}$ assuming Gaussian distributed noise [81]. The control signal data consisting of only few bytes thus can be transmitted with a very small error probability.

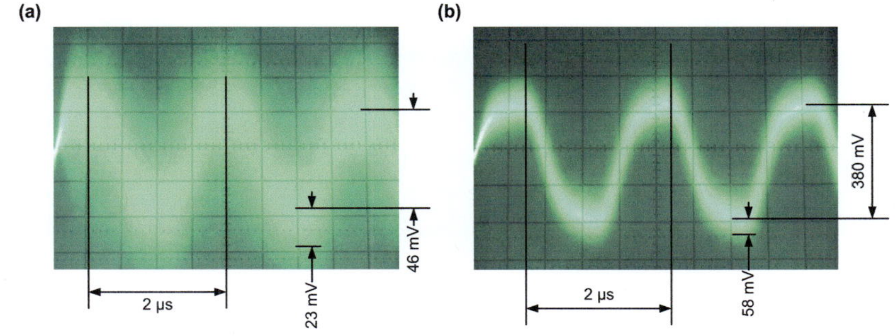

Fig. 4.6 PRBS of length $2^{31}-1$ in the downstream signal. The downstream signal was modulated with a 500 kHz square-wave signal and 10 % modulation depth. (a) At an optical input power of 8 nW, the sensitivity limit is reached. (b) With 80 nW optical input power SNR exceeds 10.

The influence of the PRBS length in the downstream signal on the control signal has been investigated further. The downstream signal lies in the GHz region but also contains power in the MHz region. The length of the PRBS affects the power distribution within the signal spectrum and hence effects the receiver performance. The results are shown in Fig. 4.7, where the effective value of the noise in dependence of the PRBS length is depicted in (a). With increasing PRBS length the downstream signal spectrum is moving into the receiver amplifier passband region becoming visible as increase in noise amplitude. This effect saturates for lengths above $2^{11}-1$.

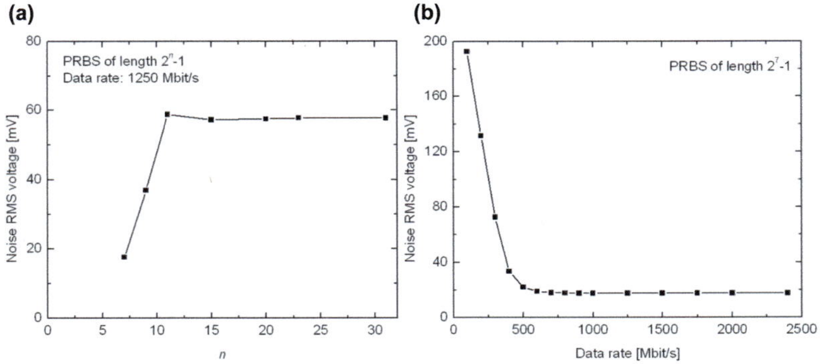

Fig. 4.7 Influence of the PRBS length on the control signal (a) With increasing length the downstream signal spectrum moves into the receiver passband region. (b) A decreasing data rate of the downstream signal has the same effect.

In the same way the data rate of the downstream signal affects the receiver performance. With increasing data rate the influence on the receiver decreases. This can be seen in Fig. 4.7 (b) where the effective value of the noise is depicted over the data rate.

Transmitter Circuit

The acquired data of the demarcation device and status reports have to be sent to the CO and are therefore modulated on a laser. The deployed data rate of 1 Mbit/s is rather slow for optical components. External modulators are not necessary, the lasers can be directly modulated meaning to imprint the data in the laser forward current. For different application scenarios two different kinds of laser were used: a vertical cavity surface emitting laser (VCSEL) with an emission wavelength of around 1550 nm [82] and distributed feedback (DFB) laser with an emission wavelength of 1490 nm [83]. The developed driver electronics for both lasers are shown in Fig. 4.8.

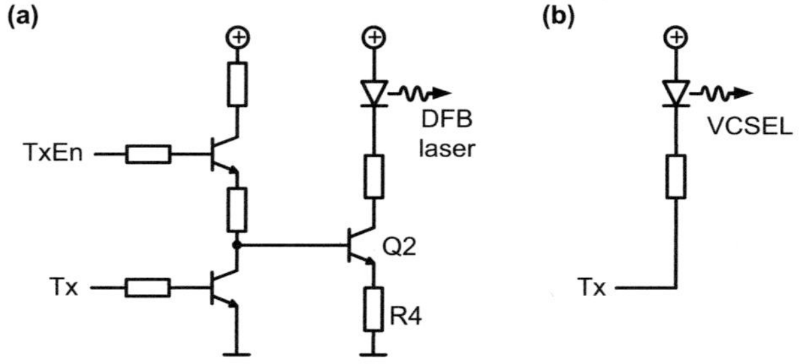

Fig. 4.8 Driver circuits for two different kinds of lasers (a) The data signal coming from the transmitter port (Tx) of the microcontroller is inverted and switches a transistor to drive current through the DFB laser. Two resistors limit the drive current and a transmitter enable signal TxEn switches the driver on or off. (b) The small currents to drive a VCSEL can be directly sourced by the data output port (Tx). The resistor limits the drive current.

The forward current to drive the VCSEL is in the range of only 7 to 9 mA and can be directly sourced by the data output port (Tx) of the microcontroller. Only an additional resistor is necessary to limit the drive current, see Fig. 4.8 (b). In contrast, the DFB laser needs drive currents in the range of 20 to 25 mA. The microcontroller is not capable to handle such currents through its data port. The driver electronic is shown in Fig. 4.8 (a). The data signal is inverted and controls the base of a bipolar transistor. Inversion is necessary as the idle state of the universal asynchronous receive transmit (UART) protocol is logic high and the laser has to be switched off in idle state to save energy. The transmitter enable port (TxEn) switches the driver on or off. The collector current through Q2 is the drive current for the laser and is limited by two resistors. The resistor R4, connected to the emitter, protects as current feedback resistor the laser from destructive overcurrent. Loss currents in the off state of the driver are kept at a minimum by resistors in the signal paths to Tx and TxEn.

Demonstrator

The above described circuits for receiver and transmitter have been brought together with the microcontroller on a printed circuit board (PCB). Power for this demonstrator of a demarcation device was drawn out of a long-life battery attached to the circuit board. Two photographs of the PCB are shown in Fig. 4.9 with the functional blocks marked.

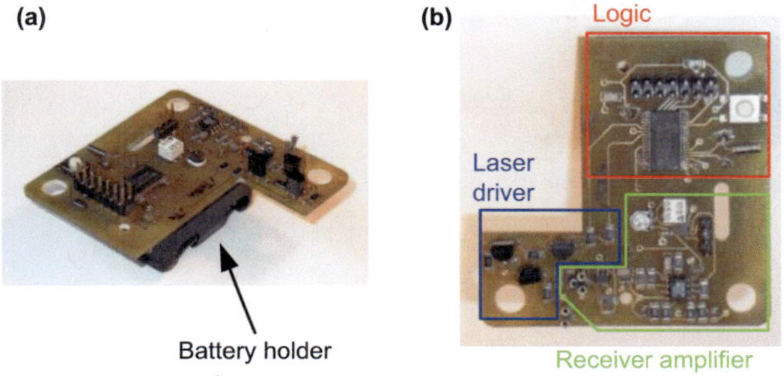

Fig. 4.9 Photographs of the printed circuit boards for a demonstrator of a demarcation device (a) The long-life battery is fixed on the bottom side. (b) The top side carries the electronic circuits for logic, laser driver (shown here for the DFB laser) and receiver.

The PCB were equipped with the opto-electronic components laser- and photodiode and integrated in a special housing. Schematics and PCB layouts for both realizations can be found in Appendix A.3. The housing was designed in a way to fit on a standard wall outlet to prove that such a demarcation device can be easily installed in every household. Two photographs of the demarcation device are shown in Fig. 4.10, where (a) shows the PCB with connected electro-optic components. The cover of PCB, Fig. 4.10 (b), carries the passive optical components like the couplers and splices. A fiber connector for the customer's network hardware is also included in the housing.

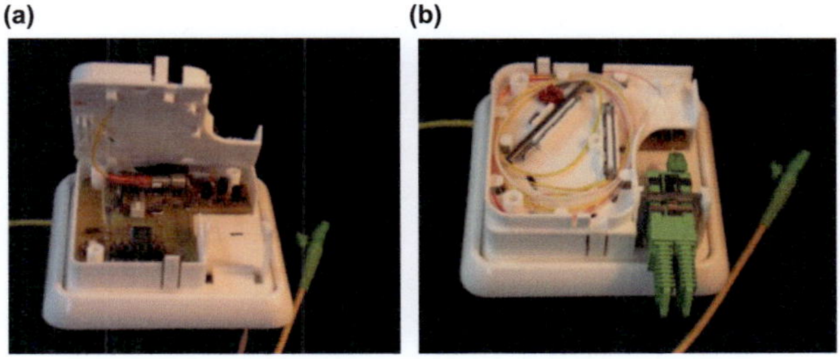

Fig. 4.10 Photographs of the demarcation device demonstrator (a) PCB with connected laser- and photodiode (b) The cover of the PCB carries the passive optical components coupler and splices. Customer's network hardware will be linked at the green E2000 connector.

Optical Power Supply for the Demarcation Device

A single photodiode (PD) illuminated with a power of $P_{\mathrm{opt}} = 5\,\mu\mathrm{W}$ at a wavelength of $\lambda_{\mathrm{pow}} = 1550\,\mathrm{nm}$ (frequency f_{pow}) delivers an open-circuit voltage $U_{oc,1}$ of about 400 mV. To increase this voltage we use N nominally identical PD connected in series. The incoming light illuminates all PD with equal powers P_{opt}/N. The single-PD open-circuit voltages $U_{oc,1}$ add up to the N-PD open-circuit voltage $U_{oc,N}$, see Eq. (2.1.7) and Eq. (2.1.17):

$$U_{oc,N} = N \cdot mU_T \ln\left(\frac{S \cdot P_{\text{opt}}}{I_s} + 1\right) - mU_T \cdot N \ln N \tag{4.1.1}$$

The short-circuit photo current of each PD is $I_{\text{sc},N} = SP_{\text{opt}}/N$. The measured current-voltage characteristics of a series connection of 4 to 8 photodiodes is shown in Fig. 4.11 for a total illumination power of $P_{\text{opt}} = 5\,\mu\text{W}\ (-23\,\text{dBm})$. We used PD with a very low saturation current $I_s = 0.3\,\text{nA}$ and a high sensitivity $S > 0.9\,\text{A/W}$ at $\lambda = 1550\,\text{nm}$.

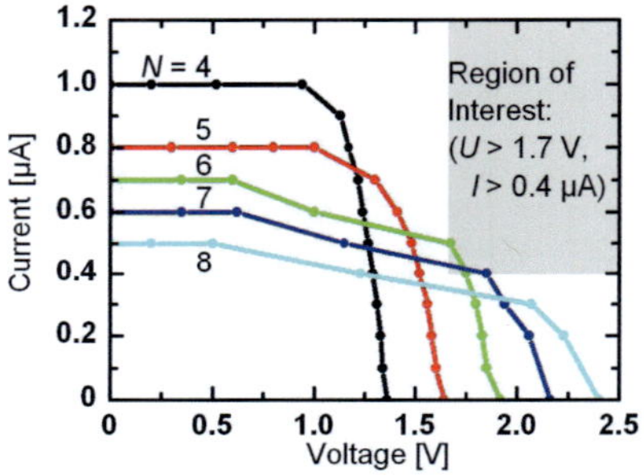

Fig. 4.11 Current-voltage characteristics of 4 to 8 PD with saturation current $I_s = 0.3$ nA in an optically parallel and electrically serial connection. Total illumination power was $P_{\text{opt}} = 5\,\mu$W. The grey-shaded rectangle marks the supply power region for the DD. Figure reprinted from [C4] © 2010 OSA.

With increasing N the output voltage increases and the output current decreases. For operating the DD a current $> 0.4\,\mu\text{A}$ and a voltage $> 1.7\,\text{V}$ are required, marked as grey-shaded rectangle. A number of $N = 6$ PD matches the DD requirements "$P_{\text{opt}} = 5\,\mu\text{W}$" optimally. Fig. 4.12 shows the single-photodiode open-circuit voltage $U_{oc,1}$ calculated from measured open-circuit voltages of a series connections of 4 to 8 photodiodes as a function of the total optical power P_{opt}. The curves virtually coincide, thereby proving Eq. (4.1.1) correct. The importance in choosing PD with low I_s is demonstrated with two additional curves in Fig. 4.12 for $I_s = 0.03\,\text{nA}, 3\,\text{nA}$. The smaller I_s, the larger $U_{oc,N}$ becomes (dashed curve), and thus the available electrical power increases.

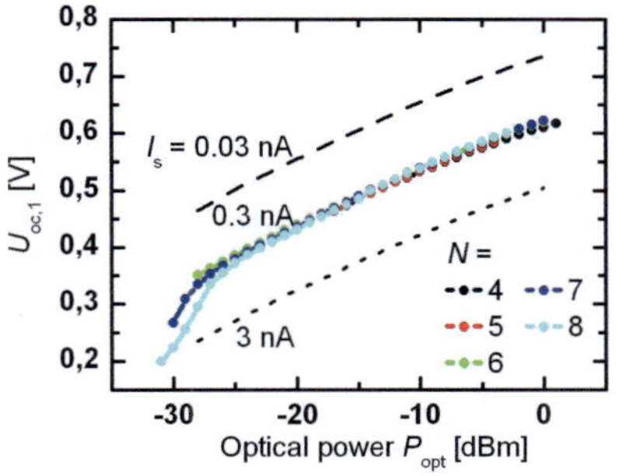

Fig. 4.12 Single photodiode open circuit voltage $U_{oc,1}$ in dependence of optical power. $U_{oc,1}$ is calculated from measured $U_{oc,N}$ assuming $mU_T = 50$ mV. For comparison, $U_{oc,1}$ is calculated for $I_s = 0.03$ nA (dashed line) and $I_s = 3$ nA (dotted line). Figure reprinted from [C4] © 2010 OSA.

Distributing the optical power on an even number of photodiodes can be done by discrete coupler-PD combinations. More advantageous, and required for an odd number, the discrete PD are replaced by sectored integrated PD. Very good results in saturation current and output voltage have been reported [54].

Different approaches for delivering the power to the demarcation devices are feasible and depend on the actual network topology. Such topologies and realizations of demarcation devices are discussed in the following subsections.

4.1.1 Access Network Scenarios

The following results will be submitted for publication:

In-service Monitoring of PON Access Networks with Powerline Independent Devices

M. Roeger, B. Hiba, J. Hehmann, M. Straub, H. Schmuck, M. Hedrich, T. Pfeiffer, C. Koos, J. Leuthold, W. Freude

4.1.1.1 Remote Powering

The necessary optical power has to be delivered through the optical distribution network (ODN). Different topologies for the network are possible. A change in topology influences the ODN loss (GPON, [55]), and affects the available power for the DD. Thus, the actual design of a DD has to be adapted to the network topology. Both, topologies and designs, will be discussed in the following subsections.

Power Broadcast and Data Broadcast

A DD in a TDM-PON must run at low supply power levels, as in a TDM-PON all signals coming from the OLT are split up among all ONTs in the network, see Fig. 4.13. Optical power provided by a laser diode (LD) at a wavelength λ_p is also split up, multiplexed along with the downlink TDM data at λ_d, and distributed over the whole network for supplying energy to the DD. The DD extracts power at λ_p and converts it with a photodiode array to a supply current for the DD. For downstream and upstream data signals at λ_d and λ_u, respectively, the DD is transparent. Only a small percentage of data power is coupled out by a 5 %-tap for receiving (Rx) control signals. Another tap allows to transmit DD (Tx) data in the reverse direction. Due to the splitting of optical supply power, the available energy at the DD is too low to perform complex measurements or to run actuators. More advanced wavelength-selective splitting techniques in the PON may overcome this limitation.

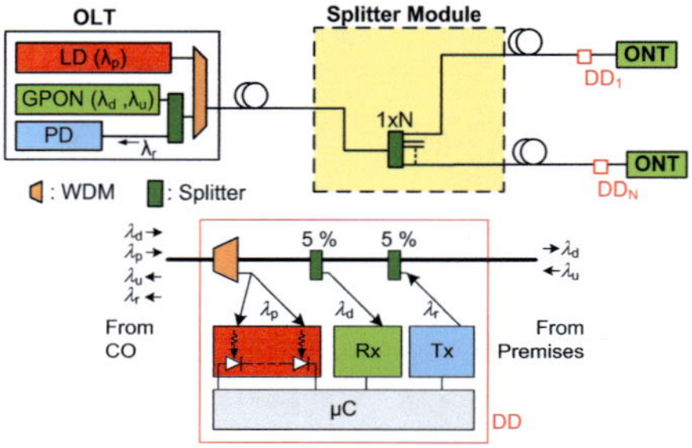

Fig. 4.13 TDM-PON with optically powered DD. One additional laser (LD) in the OLT operating at an extra wavelength λ_p provides power for all DDs within a PON. A WDM coupler couples the power at λ_p to the DD and feeds several photodiodes connected in series supplying electrical energy to the DD. The DD is transparent for upstream λ_u and downstream λ_d signals of the access network, but absorbs the power at wavelength λ_p. The DD reports its signals back on wavelength λ_r. An additional photodiode (PD) in the OLT detects the DD feedback. Basically, the GPON transceiver and the PD could be directly connected to the fiber network via the WDM coupler. However, for a better comparison with the arrangements in Fig. 4.16 and Fig. 4.22, which work without the power supplying laser LD, the splitter is kept in place.

Power Selective and Data Selective

An alternative to a special power supply laser would be to use the data signal itself for supplying energy to a DD. In a WDM-PON each ONT uses its dedicated wavelength for upstream (λ_{ui}) and downstream (λ_{di}) signaling, see Fig. 4.14. For avoiding unnecessary loss, the splitting of the feeder line to multiple ONT can be done with a cyclic arrayed waveguide grating (cAWG, [56]) having much smaller loss (6 dB for an 8-port-cAWG) than a conventional power splitter (11 dB for a 1:8 splitter). With this scheme much more power is available at the customer's premises and supplying the DD by tapping the downstream signal and using only one photodiode for power conversion becomes feasible. However, this solution is costly as for each ONT a dedicated transceiver is needed in the OLT.

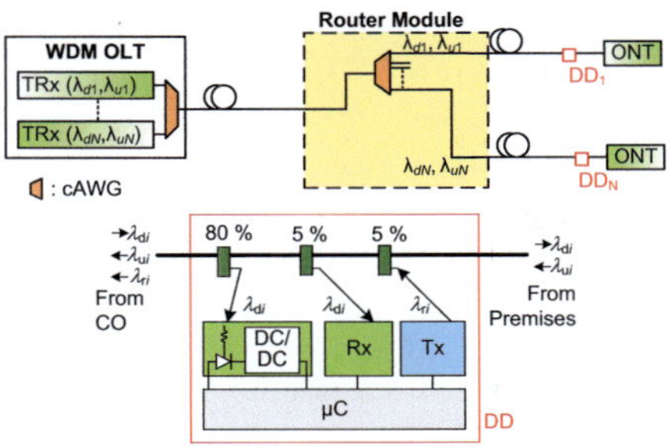

Fig. 4.14 WDM-PON with optically powered DD. In the WDM-PON much more power is available at the ONT. The DD uses 80 % of the downstream signal to supply its own operating energy. Cyclic arrayed waveguide gratings (cAWG) multiplex the up- and downstream signals in the OLT and route them in the router module to the ONTs. Each ONT is connected to its own transceiver (TRx) in the WDM OLT.

Power Selective and Data Broadcast

A combination of TDM-PON and WDM-PON can overcome both the power and cost issue. A TDM approach is used for communication between CO and ONT, and a tunable laser together with a WDM filter delivers power to each DD sequentially. To this end, the passive power splitter of Fig. 4.13 is replaced by a combination of a passive splitter and a WDM filter, see Fig. 4.15. Together with a tunable laser in the OLT selective remote powering of the DD is feasible. The optical splitter/filter module distributes the communication downstream wavelength λ_{d} to all ONTs just like a conventional passive power splitter. The same holds true for the upstream wavelength λ_{u}. Furthermore, the splitter/router module distributes the optical power of the tunable laser in the OLT (wavelength $\lambda_{\mathrm{p}n}$) to only one output port. Such splitter/router modules have recently been built in form of an integrated dual function arrayed waveguide grating [57], and as optical power combiner/wavelength demultiplexing modules in planar lightwave circuit technology [58]. On each output port one ONT is attached, with a series-connected DD for monitoring purposes. The DD, see Fig. 4.15, is supplied with light from the tunable OLT laser, whereas it is transparent for upstream and downstream signals. In a PON based on such a splitter/router module, the whole output power of the tunable laser in the OLT is delivered to one DD. All other DD in the same PON remain unpowered.

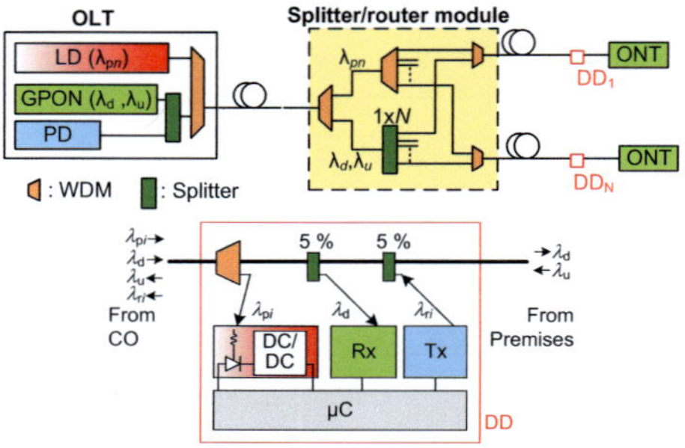

Fig. 4.15 TDM-PON: Optically powered demarcation device (DD) in a PON with WDM overlay for selective powering. Light of a tunable laser diode (LD) in the OLT (tuning range λ_1 - λ_N) is multiplexed with the conventional GPON signals into a transport fiber and connected to the splitter/router module. This module does a power splitting of the GPON signals and routes the light of the tunable laser to only one of its output ports. In the DD addressed by the specific choice of the tunable LD wavelength, this light is fully coupled out to supply the DD.

The special splitter/router module used in this selective remote powering system can be further useful for optical time domain reflectometry (OTDR) measurements. With the tunable laser as source for an OTDR system, each branch of the PON can be measured independently of the other branches avoiding ambiguous results and high losses caused by passive power splitters [59]. As all signals coming from the tunable laser are filtered by the DDs, OTDR measurements can be run in-service without affecting the data traffic.

4.1.1.2 Local Powering

Instead of supplying the power by optics, a local energy source might be used. Connecting the DD to the power grid would severely limit the range of possible applications. In addition, with a power consumption of a few µW it seems to be unjustified to connect the DD to the power grid. As an alternative, the necessary energy might be harvested from the environment. However, a solar cell would need occasional illumination, a thermoelectric generator needs a temperature gradient, and harvesters for mechanical energy require moving parts. This excludes operation in dark compartments or rooms where a DD would be most likely deployed.

However, with a long-life lithium battery (e.g. Saft, LS 14250, 1.1 Ah, 3.6 V, ½AA size [27]) as power supply for the DD shown in Fig. 4.16, the necessary power can be supplied for significantly more than 10 years. A disadvantage is the need for replacement. Advantageous is the very compact setup and the straight forward power supply design. No additional modifications in the network are necessary.

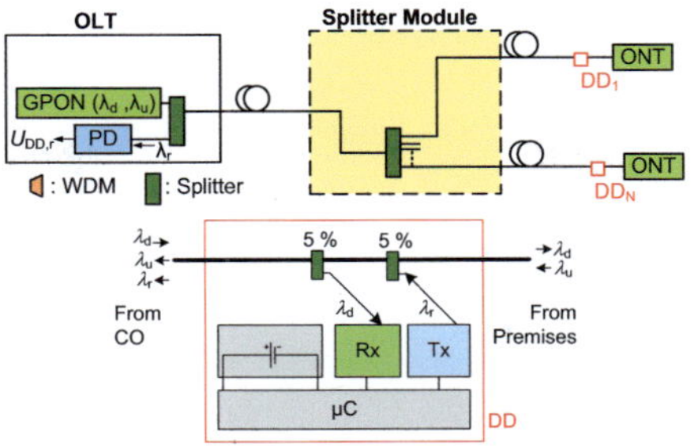

Fig. 4.16 TDM-PON with battery powered DD. No WDM components are necessary if the DD is supplied by a battery. This monitoring setup is fully compatible to standard PON techniques.

4.1.2 Access Network Measurements

The following results will be submitted for publication:

In-service Monitoring of PON Access Networks with Powerline Independent Devices

M. Roeger, B. Hiba, J. Hehmann, M. Straub, H. Schmuck, M. Hedrich, T. Pfeiffer, C. Koos, J. Leuthold, W. Freude

In this section we discuss the implementation of the various demarcation devices. We begin the section with discussing how to realize the power converter, and then continue to show the implementations of the remote and locally powered DDs.

Efficiency of Opto-Electronic and DC/DC Boost Converters

To power the DD we require a minimum supply in the µW range. This power can be provided optically. However, a photovoltaic power converter typically does not provide the minimum supply voltage of 1.7 V needed for electronics. We therefore either need a photodiode array operating optically in parallel and connected electrically in series, or a DC/DC booster must be employed. We will subsequently discuss both options.

The optical power for the DD can either be derived from the optical data signal, or from an additional laser in the OLT. The laser could operate at an extra wavelength and must provide sufficient power for all DDs within a standard PON. Assuming a maximum ODN loss of 30 dB (GPON, Class C [55]) and an output power of the extra laser of +8 dBm, a total of –22 dBm ($= P_{\text{opt}} = 6.3\,\mu\text{W}$) is available for powering a DD.

The open circuit voltage provided by a single photodiode (PD) illuminated with P_{opt} is [18], see also (2.1.7),

$$U_{\text{oc},1}(P_{\text{opt}}) = mU_{\text{T}} \ln\left(\frac{SP_{\text{opt}}}{I_{\text{s}}} + 1\right). \tag{4.1.2}$$

This gives 400 mV for a PD illuminated at a wavelength of $\lambda_{\text{pow}} = 1550\,\text{nm}$ (frequency f_{pow}) with a PD sensitivity $S = \eta e / (hf_{\text{pow}}) > 0.9\,\text{A/W}$ (quantum efficiency η, Planck's constant h), saturation current $I_s = 0.3\,\text{nA}$, ideality factor m and the thermal voltage $U_T = kT/e$ (Boltzmann constant k, temperature T, elementary charge e). We used PDs with a very low saturation current I_s and a high sensitivity S at $\lambda_{pow} = 1550\,\text{nm}$. The short circuit current $I_{sc,1}$, see also (2.1.6),

$$I_{sc,1}(P_{opt}) = SP_{opt}, \tag{4.1.3}$$

which is in the order of 6 µA for $P_{opt} = 6.3\,\mu\text{W}$.

An efficient way to increase the voltage is to use N nominally identical PDs connected in series. The incoming light is split up and illuminates all PDs with equal powers P_{opt}/N. By connecting N PDs in series, the single-PD open-circuit voltages $U_{oc,1}$ increases to $U_{oc,N}$, whereas the short-circuit photocurrent of the PD array $I_{sc,N}$ decreases,

$$\begin{aligned} &U_{oc,N}(P_{opt}) = NU_{oc,1}\left(\frac{P_{opt}}{N}\right) \approx NU_{oc,1}(P_{opt}) - mU_T \cdot N \ln N, \\ &I_{sc,N}(P_{opt}) = I_{sc,1}\left(\frac{P_{opt}}{N}\right) = \frac{1}{N} I_{sc,1}(P_{opt}), \\ &SP_{opt} >> I_s. \end{aligned} \tag{4.1.4}$$

The measured current-voltage characteristics of a series connection of 4 to 8 photodiodes is shown in Fig. 4.17 for a total illumination power of $P_{opt} = 6.3\,\mu\text{W}\ (-22\,\text{dBm})$. When increasing N, the output open circuit voltage increases, and the output short circuit current decreases. For operating the microcontroller of the DD, a minimum current of 0.55 µA and a minimum voltage of 1.7 V are required. This operation region is marked by a grey-shaded rectangle in Fig. 4.17 (a). It can be seen that for $N = 6$ or $N = 7$, the PDs generate sufficient current and voltage to supply the DD. In Fig. 4.17 (b) the measured electrical output power and the conversion efficiency of the photodiode arrays are shown. When increasing the number of photodiodes, the maximum available output power decreases. This results from the logarithmic voltage drop when decreasing the power incident onto a photodiode. The efficiency drops from 22 % for an array of 4 PDs to 14 % for 8 PDs. Not shown in the graph is the conversion efficiency of a single photodiode which has been measured to be 30 %.

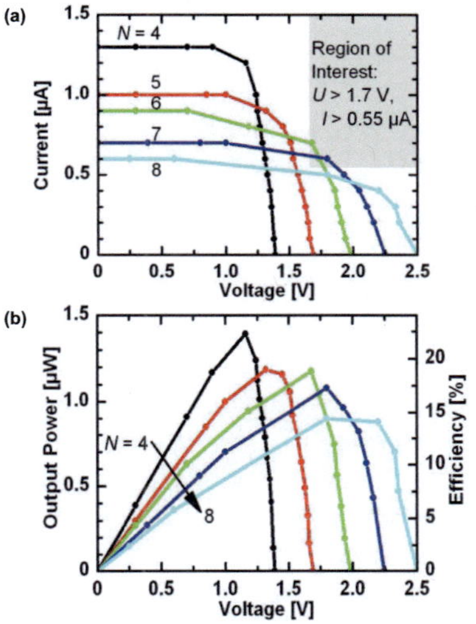

Fig. 4.17 Output current and power vs. output voltage of arrays with 4 to 8 photodiodes illuminated in parallel with a total optical power of 6.3 µW, and connected electrically in series. (a) With the number of photodiodes the open circuit voltage increases, whereas the short circuit current decreases. The grey shaded region marks currents and voltages needed to run the µC electronics. (b) Electrical output power and conversion efficiency from different PD arrays. When increasing the number of photodiodes, the maximum output decreases and conversion efficiency drops from 22 % to 14 %.

Another way to increase the output voltage is by using a DC/DC boost converter circuit. DC/DC boost converters are commonly used for upconversion of low DC voltages to higher DC supply voltages. Unfortunately, they are extremely inefficient for conversion of ultra-small powers.

The network scenarios discussed in 4.1.1 show that we must be able to handle power levels of as little as a few µW. In Table 4.1 power efficiencies for two low power DC/DC converters (TPS61201 from Texas Instruments and LTC3105 from Linear Technology) at different operation points are collected from the converter datasheets. These devices are capable to boost the output voltage of a single photodiode efficiently from 400 mV to 2 V. However, efficiency strongly depends on the input voltage and the attached load. With decreasing load, the efficiency drops significantly. The minimum output current specified for TPS61201 is 0.1 mA where efficiency already drops to 35 %. For LTC3015, an efficiency of 28 % results at an output current of 0.02 mA. Decreasing the output current even further would result in a drastic loss of efficiency. Such poor conversion efficiency would apply for the times where the DD is in sleep mode, thereby consuming less than 1 µA. As a consequence, energy accumulation in a storage device would not be possible. Note that the efficiency of the booster has to be multiplied with the efficiency of the single photodiode (30 %) to get the overall efficiency.

Table 4.1 Efficiencies of DC/DC boost converters

Operation	(V_{out} = 3.3 V all)	TPS61201 [60]	LTC3105 [61]
Optimum	(V_{in} = 1.0 V, I_{out} = 10 mA)	62 %	80 %
Low output power	(V_{in} = 1.0 V, I_{out} = 0.1 mA)	35 %	58 %
Low input voltage	(V_{in} = 0.6 V, I_{out} = 10 mA)	63 %	65 %
Low input power	(V_{in} = 0.6 V, I_{out} = 0.02 mA)	–	28 %

Thus, for upconverting the voltage resulting from small available optical powers, the method of using several photodiodes connected electrically in series is more efficient. However, it is more expensive as a larger number or more complex optical components are needed. If the available optical power is high, a single photodiode in combination with a DC/DC boost converter can be used for supplying the DD.

4.1.2.1 Remote Powering

In this section we discuss results from implementations of the power broadcasting approach, the power-selective (both discussed in 4.1.1.1), and the battery powered DD approaches (discussed in 4.1.1.2).

Power Broadcast and Data Broadcast

At the beginning of Section 4.1.2 it has already been shown that an optical power $P_{opt} = 6.3\,\mu W$ illuminating an electrical series connection of 6 photodiodes results in a photocurrent and a voltage that is sufficient to run the DD in low-power mode.

Another important fact is the initial startup procedure which has to be considered in the design. In the initial state of the DD the storage capacitor C_S is discharged and shows zero voltage, see Fig. 4.18. The illuminated photodiodes start to charge the capacitor so that its voltage slowly increases. A microcontroller (µC) is connected to C_S and operates with a voltage below the specified minimum (here 1.7 V), therefore runs in an undefined operating mode and can consume a current of up to several milliampere. So, during the initial charging period when the capacitor voltage increases from 0 to 1.7 V, the microcontroller has to stay disconnected. Connecting µC and C_S has to be done by an extra circuit which supervises the voltage at C_S. Such voltage supervisor ICs are readily available. Here, a LTC2935 [62] was used which has a selectable switching voltage and excels with a very low current consumption of only 500 nA at 2 V. The output port switches to high if the input exceeds a given threshold. An inbuilt hysteresis avoids instable states. The initial start procedure of the sensor node is then as follows: The voltage supervisor is connected to the storage capacitor whereas the microcontroller is not, see Fig. 4.18 (a). If the voltage at C_S has reached the minimum voltage of 1.7 V needed for running the microcontroller, the voltage supervisor connects the microcontroller to C_S. The microcontroller in turn disconnects the supervisor to keep power consumption at a minimum, see Fig. 4.18 (b). The power management is now fully done by the microcontroller. The surplus of energy after the start-up period is stored in C_S, and the DD can decide on its different operating modes.

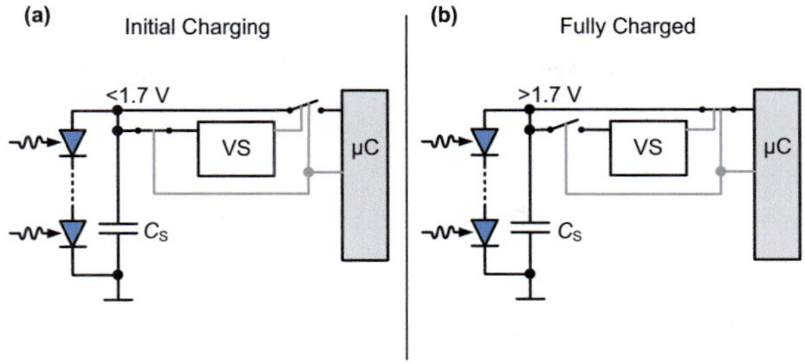

Fig. 4.18 Switching scheme in the initial startup procedure (a) The necessary microcontroller (µC) operating voltage of 1.7 V is not yet reached, a voltage supervisor (VS) is connected to an energy storage capacitor C_S. (b) When the voltage at C_S reaches 1.7 V, the VS connects C_S and µC, which in turn disconnects the VS.

Power Selective and Data Broadcast

In power selective schemes much more power is available at the DD site, and more complex sensing applications become feasible. Even actuators could be operated. For demonstration purposes a prototype DD has been built which connects and disconnects an ONT physically from the network. The DD features a micro-electromechanical system (MEMS) switch, see Fig. 4.19.

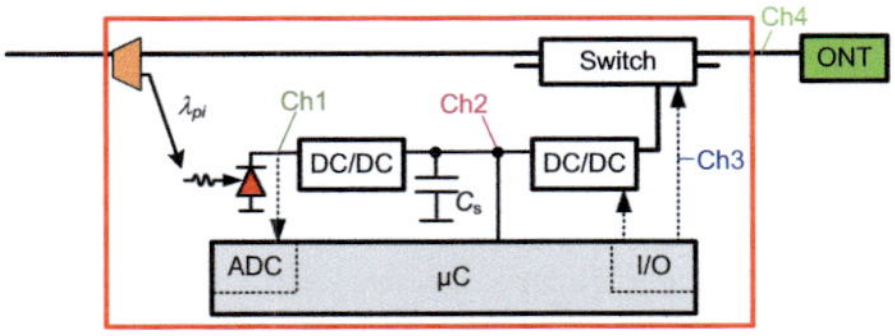

Fig. 4.19 Block diagram of the optically powered DD. The optical supply channel ($P = 3.5$ dBm) is tapped by a WDM coupler and fed to a photodiode. Its output voltage is boosted by a first DC/DC converter to 2.2 V supplying a microcontroller (µC) and charging a capacitor C_S. Control signals sent from OLT to DD are encoded in the supply power duration. To receive the control signals, the µC monitors with its analog-to-digital converter (ADC) the output voltage of the PD and detects whether the supply channel is switched on or off. One important control signal serves to disconnect the ONT from the network with a built-in switch. The necessary switching energy is also taken from C_S. The capacitor's voltage is boosted by another DC/DC converter. Both, converter and switch, are controlled by general input/output ports (I/O) of the µC. Ch1 to Ch4 mark measurement points with signals shown in Fig. 4.20.

To demonstrate this feature, we operate the DD with a wavelength of $\lambda_{pi} = 1550\,\text{nm}$ and an optical input power of 3.5 dBm using a tunable external cavity laser. This relatively large amount of input power when detected with a single photodiode suffices to drive a DC/DC boost converter which provides the necessary voltage for running the DD's electronics. At several measurement points the voltages were recorded and are plotted in Fig. 4.20 as a function of time, illustrating the operating principle of the DD. The photodiode in the DD delivers a voltage of 0.5 V (Ch1, yellow) when illuminated with an optical power of 3.5 dBm. The PD output voltage is boosted by a DC/DC converter to supply the µC with a voltage of 2.2 V (Ch2, red). After sensing that the supply channel has been switched off, the µC processes the control signal "connect" (Ch3, blue). Here, the control signal is encoded in the supply duration of the DD. Additional addressing signals are not necessary as for the "selective-power" scheme under discussion only a single DD is powered at a time. Subsequently, a second DC/DC boost converter is started to supply the MEMS switch, which in turn changes its switching state (Ch4, green), and disconnects (or connects in this case) the attached ONT and the network.

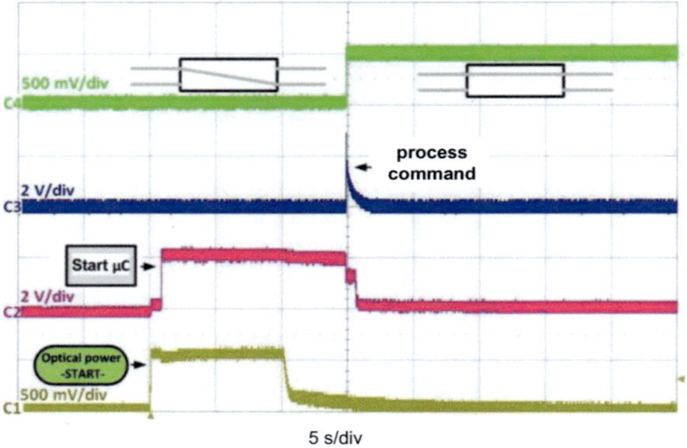

Fig. 4.20 Connecting an ONT to the network via the DD of Fig. 4.19. The photodiode delivers a voltage of 0.5 V (Ch1, yellow) under illumination with an optical power of 3.5 dBm. This voltage is boosted by a DC/DC converter to supply the µC with a voltage of 2.2 V (Ch2, red). After sensing that the supply channel has been switched off, the µC processes this control signal (Ch3, blue). As a consequence a second DC/DC boost converter starts supplying power to the MEMS switch, which in turn changes its switching state (C4, green) thus disconnecting (or connecting in this case) the attached ONT to the network.

4.1.2.2 Local Powering

As a second possibility, a battery powered DD can be placed close to the customer's premises. We developed a prototype of such a device. Its exploded view is shown in Fig. 4.21. It consists of a housing which can be attached to a standard flush-mounted wall outlet. The fiber coming from the optical distribution network (ODN) passes the wall outlet and is connected with a splice to the fibers in the DD. The optical and electrical components are placed separately in different planes of the housing. The customer's ONT can be connected with a standard E2000 connector at the DD to the PON.

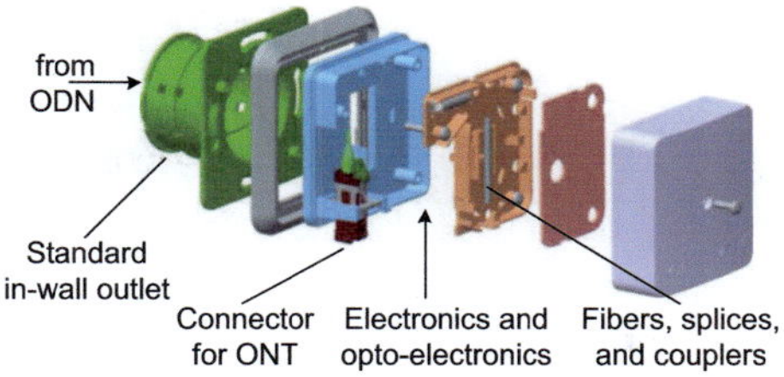

Fig. 4.21 Exploded view of a housing for a battery powered DD. The optical components like fibers, splices, and couplers are separated from the electronics (not shown) in the housing. The housing is attached to a standard flush-mounted wall outlet. The housing also includes an optical standard E2000 connector for attaching the ONT.

Three battery powered prototypes were successfully integrated and tested in a GPON testbed shown in Fig. 4.22. The testbed consists of one optical line termination (OLT) transceiver which was upgraded to communicate with the DD. The downstream GPON laser sends control signals to the DD as a small perturbation (1 Mbit/s NRZ, modulation depth 10 %) of the GPON signal. This perturbation does not yet cause frame loss at the ONT, as it only occupies a small part of the downstream electrical signal spectrum [78]. An additional photodiode (PD) receiver in the OLT detects the control signals received from the DD and converts them to an inverted, electrical signal $U_{\mathrm{DD},r}$. The OLT connects via several splitters and fibers of different length to two optical network terminations (ONT). Each ONT is connected through a DD to the network. Standard GPON traffic with a pseudo-random bit sequence of length 2^7-1 as payload was transmitted from the OLT to the ONT, simultaneously polling the DD. The polling procedure showed no effect on the GPON transmission.

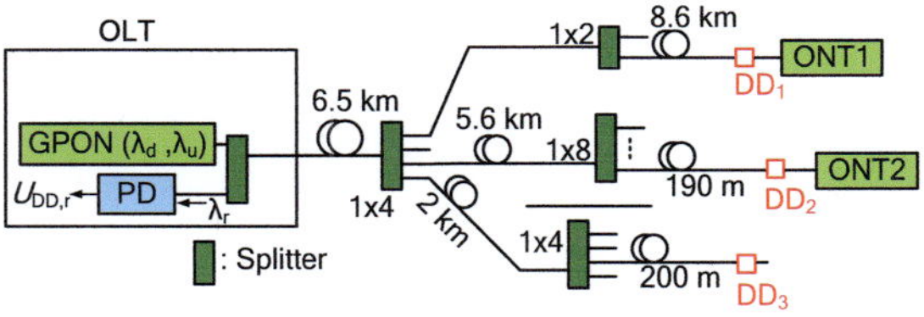

Fig. 4.22 Setup drawing of the GPON testbed. The OLT transceiver polls the three integrated DD while handling GPON traffic with two attached ONTs. Payload of the GPON was a pseudo-random bit sequence with a length of 2^7-1. Different splitting ratios and fiber lengths are used to ensure a realistic application scenario. The response signals coming from the DDs are detected with an additional photodiode (PD) in the OLT transceiver and converted into an inverted, electrical signal $U_{\mathrm{DD},r}$.

In Fig. 4.23 the result of the test in the network is shown. A LabView interface is used for controlling the OLT. The measured received signals $U_{\mathrm{DD},r}$ from the DD are read out and plotted via the LabView interface, a screenshot of which is shown in Fig. 4.23. The inverted response signals can be seen as downward spikes in the trace. The different power levels originate from different path losses for the different DD in the exemplary PON. An inset enlarges the right-most DD signal and reveals the single bits of the NRZ response.

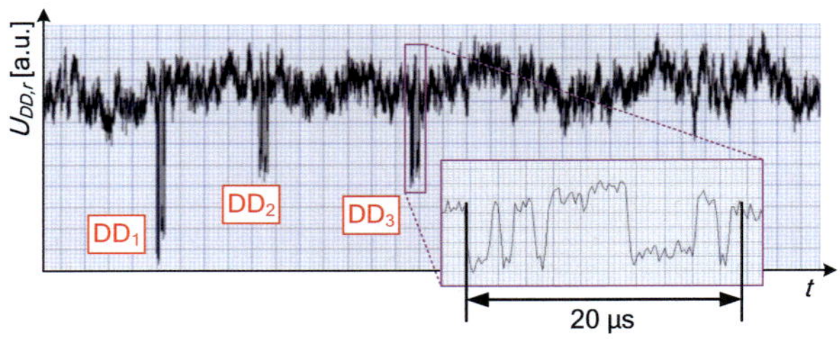

Fig. 4.23 Inverted receiver output voltage $U_{\mathrm{DD},r}$ showing measured response signals of three polled DD in the GPON testbed of Fig. 4.22. The OLT requests the DD subsequently to send a response. The three groups of downward spikes correspond to the signals of the DD. The different power levels originate from different path losses in the exemplary PON. In the box one NRZ response signal is enlarged. It consists of 20 bits at a bitrate of 1 Mbit/s.

Summary

We proposed and demonstrated a new concept for monitoring passive optical networks where OTDR techniques are not feasible. The optically or battery powered monitoring DDs are composed of standard low-cost components. Supplying power to the off-grid DDs was identified as the most critical task, and feasible DD designs for different topologies of the access network were discussed. As exemplary monitor functionality, we implemented an availability test for a link associated unambiguously with the customer's site. In addition, we demonstrated that a customer's network access can be enabled or disabled remotely by the CO. Other monitoring functions — like measuring optical power levels at the customer's site or the DD's battery status — are easily implemented in the electronic part of the DD. Polling of the DD and information requests can be run in-service using our new LE-MAC protocol, and does not affect GPON traffic. This easy-to-implement and inexpensive design is a promising candidate for monitoring future access networks.

4.2 Sensor Devices in Fiber Networks

The results of this section were partially published in [C1] and studied in two Bachelor Theses [S7], [S8].

Energy is a precious resource, and telecommunication networks tend to demand more of it due to the fast growing traffic volume, so energy efficiency becomes an important issue [84]. Much energy is wasted by devices which are fully powered up without having traffic to deal with. The easiest and most efficient way to save energy is to switch off network devices or components which are currently not in use [85]. Because switched-off, "sleeping" devices are not listening to requests and cannot be addressed; appropriate protocols need to re-establish the communication channel first.

Here, we describe a protocol which handles the communication with continuously active and with temporarily sleeping network elements [72]. For low duty-cycle sensors, sleep modes can be exploited efficiently, so that the average power consumption can be decreased to a level where even optically powering of network elements becomes feasible. As an example, we demonstrate a long-range optical network of sensors, where only 16 μW optical power per sensor node is sufficient for its operation. Such networks are of interest if galvanic connections must be avoided, or if a local electrical power supply does not seem appropriate.

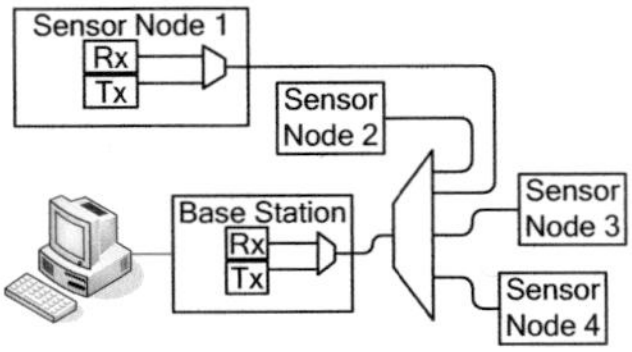

Fig. 4.24 Example network with base station (BS) and sensor nodes S*i*. Nodes communicate only with BS. Figure reprinted from [C1] © 2011 OSA

In an experimental setup with electrically connected sensor nodes, Fig. 4.24we checked the operation of the LE-MAC protocol. An ARM STR-E912 evaluation board acted as base station, four boards equipped with MSP430-F2234 μC represented sensor nodes [36]. The boards where directly connected over their UART ports. The effective co-existence of sensor nodes with high availability (duty cycle 1) and low availability (duty cycle 3×10^{-5}) within one network was successfully demonstrated. Settings in the exemplary network were:

- Data rate: 1 Mbit/s
- T_{Lstn}: 2 ms
- T_{Com}: 5…10 ms
- T_{Sleep}: 60 s
- Packet size: 30 byte

4.2.1 Optically Powered Sensor Network

When a protocol can handle loss of synchronism, so that network elements can employ lowest-power sleep modes, even optically supplied power can suffice to operate a sensor node. In Fig. 4.25 (a), an optically powered sensor network is sketched. The base station (BS), responsible for data collection and energy supply, comprises a standard DFB laser ($\lambda = 1510\,\mathrm{nm}$, $\lambda_{BS,\max} = +6\,\mathrm{dBm}$). The laser operates continuously (CW) and provides the nodes with the necessary optical power. The CW light is slightly modulated (modulation depth 10%) for transmitting the necessary protocol data to the sensor nodes. Three sensor nodes are connected to the base station via an optical 1:8 splitter. One of the nodes is remotely connected via a 10 km standard singlemode fiber (SMF28). The data collected by the BS are read with a standard USB connection and visualized on a PC. The USB connection also supplies electrical power to the base station, so that the whole sensor network is powered and controlled by one USB port.

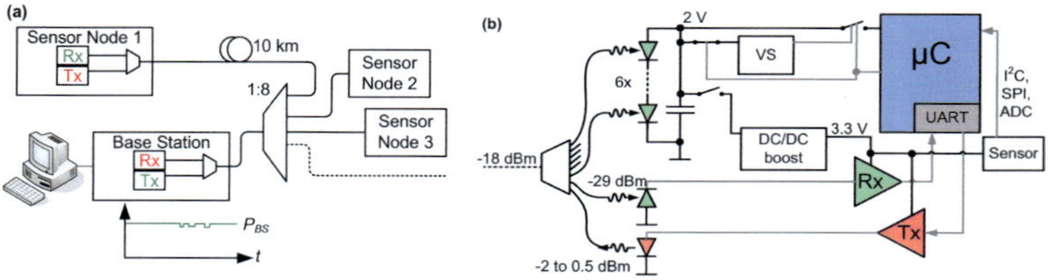

Fig. 4.25 Optically powered sensor network (a) Drawing of the sensor network consisting of a base station electrically connected to a PC and optically connected to three sensor nodes via a 1:8 splitter. (b) Block diagram of a sensor node with opto-electronic power conversion, multiple communication interfaces and power management functionalities. Figure reprinted from [C1] © 2011 OSA

In Fig. 4.25 (b) a block diagram of an optically powered sensor node is shown. The incoming light with a power of 16 μW (–18 dBm) is distributed by a 1:8 splitter to 6+1 photodiodes [54]. The upper six photodiodes supply electrical energy to the node. Each photodiode provides a voltage of 0.35 V generated out of –29 dBm optical power. As the photodiodes are connected in series, they deliver up to 2 V for the microcontroller (μC). A 470 μF capacitor acts as energy storage, and is charged by the photodiode array. In sleep and snooze mode the surplus of electrical power coming from the photodiode array charges the storage capacitor, whereas in active periods the capacitor is partially discharged by 0.1 V. For one S*i* in sleep mode, the current consumption was measured to be 550 nA. Recharging the capacitor after an active period of 1.5 ms takes 8 min. The voltage at the capacitor slowly increases with time. However, the μC does not reliably operate at startup, where supply voltages are very low and only slowly increasing, therefore an auxiliary device, a so-called voltage supervisor (VS) is needed. When the voltage at the storage capacitor reaches 2 V, the VS switches the μC power on. Then the μC switches the VS off. The complete power management now remains with the μC. If the sensor node needs to connect to the base station, a DC/DC boost converter is started and delivers (for a limited time period) a fixed output voltage of 3.3 V supplying the receiver circuitry (Rx). The Rx photodiode is connected to the seventh port of the 1:8 splitter and receives the BS control signals. Sensor data are transmitted back to the base station with a directly modulated VCSEL diode (Tx, $\lambda = 1310\,\text{nm}$, $P_{\text{out}} = -2 \ldots +0.5\,\text{dBm}$). It is connected to the eighth port of the splitter. Both Rx and Tx use the UART port of the µC.

The open circuit voltage of the photodiode defines the maximum charge voltage for the storage capacitor and increases logarithmically with the optical input power. The sensitivity of the receiver circuitry is therefore influenced twofold by the input power: Directly by the optical control signal power, and indirectly because Rx is supplied by the DC/DC boost converter which is losing stability with a decrease of optical input power, resulting in a decreased input voltage for the DC/DC converter. Fig. 4.26 displays the measured frame error ratio (FER) in dependence of the optical input power. The strong increase of the FER for $P_{\text{opt}} < -21.7\,\text{dBm}$ results from instable operation of the DC/DC boost converter. As the node is normally supplied with an optical power of –18 dBm, a margin of 3 dB is left. Input powers of more than -13.8 dBm drive the receiver amplifier into saturation, and strong signal distortion occurs. In between, the FER drops below $4 \cdot 10^{-6}$.

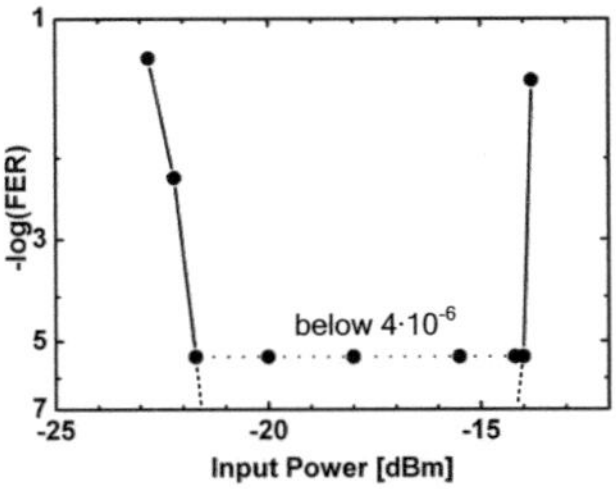

Fig. 4.26 Measured frame error ratio depending on the optical input power. Figure reprinted from [C1] © 2011 OSA

The sensor node can be equipped with a number of different sensors. The node provides 3.3 V supply voltage as well as digital (I²C, SPI) and analog sensor interfaces. In our network, the three nodes where equipped with a temperature, an acceleration and a light sensor, respectively. With a minimum input power of –18 dBm per sensor node, 24 dB are available as distribution loss. This corresponds to a maximum splitting of 1:128 including 2 dB excess loss, or up to 120 km of standard SMF.

4.2.2 Low-power Electronic Circuits

The whole sensor node is not operated continuously as many kind of sensor information are only polled once within, e. g., 10 minutes. Thus, the node is only active for a short period of time to gather data, receive commands from its base station and to send acquired and requested data back. For simple sensors this period can last only few milliseconds whereas the rest of the time the node stores the energy delivered via the supply channel. The sleep mode current consumption of the microcontroller dominates the need and is significantly smaller than the maximum current of 1.4 μA delivered by the photodiode array.

The whole node operates most efficiently if in sleep mode unused components like sensors, receiver and transmitter are disconnected from the power supply. The necessary electronic switches and further electronic components for an energy-efficient node are discussed in the following paragraphs.

Voltage Supervisor

In the initial state of the sensor node the storage capacitor is discharged and shows a voltage of zero. The illuminated photodiodes start to charge the capacitor and so that its voltage slowly increases. A microcontroller operated with a voltage below the specified minimum (here 1.8 V) runs in an undefined mode and can consume up to several milliampere. During the initial time where the storage capacitor voltage increases from 0 to 1.8 V, the microcontroller has to be disconnected. Connecting the microcontroller to the storage has to be done by another circuit which supervises the charge voltage. Exploiting a breakthrough in a Zener diode would not result in the needed abrupt transition. Specially for this purpose voltage supervisor ICs are designed. Here, a LTC2935 [62] was used which has a selectable switching voltage and stands out with a very low current consumption of only 500 nA at 2 V. The photodiode array delivers a maximum of 1.4 μA. 900 nA are left for charging the capacitor and for compensating losses. The output port switches digitally to high if the input exceeds the given threshold. An inbuilt hysteresis avoids instable states.

The initial start procedure of the sensor node is than as follows. The voltage supervisor is connected to the storage whereas the microcontroller is not. If the storage has reached the minimum voltage needed for running the microcontroller, the voltage supervisor connects the microcontroller to the storage. The microcontroller in turn disconnects the supervisor to keep power consumption at a minimum. The power management is now fully done by the microcontroller.

Switches

The key component in the power management is the switch to connect and disconnect parts of the circuit to the supply voltage. The switch may be placed between supply and load, a so-called highside switch, or between load and ground, a so-called lowside switch.

NPN transistors as common-emitter amplifier are good for implementing lowside switches, but need to be placed to each ground connection which rules out usage in complex circuits. Therefore highside switches were implemented for controlling the power supply of the microcontroller and the DC/DC boost converter.

A switch needs to have a low on-resistance and a high off-resistance. A FET might offer both, but only if driven with enough voltage. As only voltages in the range of some volt are available, a low on-resistance cannot be achieved. Off-resistances in the range of 100 GΩ are feasible. A bipolar transistor can be driven with voltages around 0.7 V and has a low on-resistance. A PNP transistor as highside switch on a positive voltage supply leads to a voltage drop of only 50 mV in the supply path. Thus, the implemented highside switches were set up of a bipolar transistor and a FET: the bipolar transistor to control the supply path with only little losses and the FET to block the bipolar effectively.

During the start-up process the microcontroller is unpowered. All its output ports are then at ground potential. All power switches therefore have to be in the off-state to avoid unwanted loading of the storage capacitor. This defines the switch polarity: ground at the switch's input has to result in the off-state.

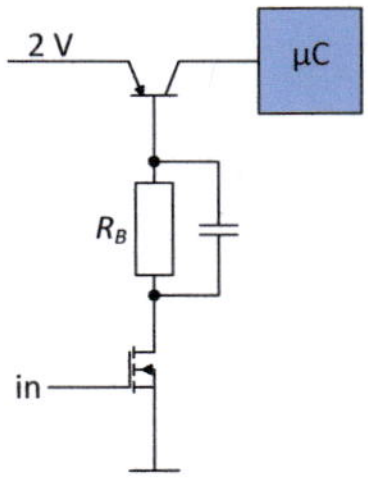

Fig. 4.27 Switch consisting of a bipolar junction and a field effect transistor (FET). The bipolar component switches the supply current and shows little losses even if operated with small voltages. The FET shows very high impedance in the off state minimizing the leakage current through the switch in the off state.

In Fig. 4.27 a schematic of the implemented switch is shown. A high signal at the input (in) makes the FET conducting and enables a base current in the bipolar to flow. The base current is limited by the resistor R_B. The base current I_B has to be large enough to enable the necessary collector current I_C to the microcontroller to flow. The current consumption of the microcontroller varies between 1 µA in sleep and 1.6 mA in active mode. The current amplification $\beta = I_C / I_B$ of the employed transistor is 500 [86]. In active mode the microcontroller needs a supply current $I_{\mu\text{C,active}}$ of 1.6 mA [36]. The maximum resistance $R_{\text{B,active}}$ for supplying the active microcontroller can be calculated to

$$R_{\text{B,active}} = \frac{2\,\text{V} - 0.7\,\text{V}}{I_{B,\text{active}}} = \frac{2\,\text{V} - 0.7\,\text{V}}{I_{\mu\text{C,active}} / \beta} \leq 406\,\text{k}\Omega\,. \tag{4.1.5}$$

The base current would then be $I_{B,\text{active}} = I_{\mu\text{C,active}} / \beta = 3.2\,\mu\text{A}$, which exceeds the current consumption of the microcontroller in sleep mode $I_{\mu\text{C,sleep}} = 1\,\mu\text{A}$. Therefore another switch is needed for maintaining power connection in sleep. For this switch the maximum resistance at the base results to

$$R_{\text{B,sleep}} = \frac{2\,\text{V} - 0.7\,\text{V}}{I_{B,\text{sleep}}} = \frac{2\,\text{V} - 0.7\,\text{V}}{I_{\mu\text{C,sleep}} / \beta} \leq 650\,\text{M}\Omega\,. \tag{4.1.6}$$

In both cases the resistance of the open FET was neglected with respect to the resistor R_B and the voltage drop over the base emitter junction in the bipolar transistor was assumed to be 0.7 V. In the implementation the resistances were chosen to $R_{\text{B,active}} = 500\,\text{k}\Omega$ and $R_{\text{B,sleep}} = 25\,\text{M}\Omega$.

The base current represents a loss. In active mode, a current of $I_{B,\text{active}} = 1.3\,\text{V} / R_{\text{B,active}} = 2.6\,\mu\text{A}$ is lost. The active microcontroller consumes 1.6 mA, outnumbering the base current loss by a factor of more than 600. In sleep mode, a current of $I_{B,\text{sleep}} = 1.3\ \text{V} / R_{\text{sleep}} = 52\ \text{nA}$ has to be compared with $I_{\mu\text{C,sleep}} = 1\ \mu\text{A}$ resulting in an efficiency of 95 %.

A low signal at the input in Fig. 4.27 pulls the resistance of the FET in the $\text{G}\Omega$ range and the base current is blocked [87]. The bipolar transistor is then also blocked, and the switch is in off-state.

An additional small capacitor with a capacitance in the pF range, in parallel to R_{B}, decreases switching time to less than 1 ns.

The combined PNP FET switch is also used to disconnect the DC/DC boost converter from the storage capacitor. The converter supplies the receiver circuit, the transmitter and sensors. Several 10 mA are expected so the base resistor for this switch was set to $20\ \text{k}\Omega$.

The switch Q1 in Fig. 4.28 in front of the microcontroller is connected to different base resistors. Depending on the actual operation mode the optimum base current can be chosen. This is needed during the start-up sequence.

Start Up Sequence

At the initial start the energy storage capacitor C_S is empty and all voltages in the system are zero. After starting the optical power supply, the storage capacitor is slowly charged and its voltage rises. To avoid unwanted loading of the storage all circuits but the voltage supervisor are disconnected with switches Q1 and Q2, see Fig. 4.28. If the voltage supervisor (VS) detects the minimum voltage of 1.8 V its output switches from 0 to 1.3 V. FET Q3 opens and the PNP transistor Q1 connects the microcontroller to the power source. The small resistance R_{active} at Q3 allows active operation of the microcontroller at 8 MHz clock speed. After the initialization the microcontroller switches Q4 to high. The microcontroller now keeps its supply path open and the voltage supervisor is disconnected from the power supply setting the base potential of Q2 high. Having finished all assigned operations, the microcontroller is ready for energy-saving sleep mode. To save energy in the switch itself Q4, is blocked setting its gate potential to ground and Q5's gate potential to high. Q5 is connected to $R_{\text{B,sleep}}$ allowing enough current to supply the microcontroller in sleep mode. After the sleep period the microcontroller switches autonomously in active mode. To cover the higher demand on power, Q4 is opened again during active times. The input of the microcontroller is additionally buffered with a 200 nF capacitor C_B.

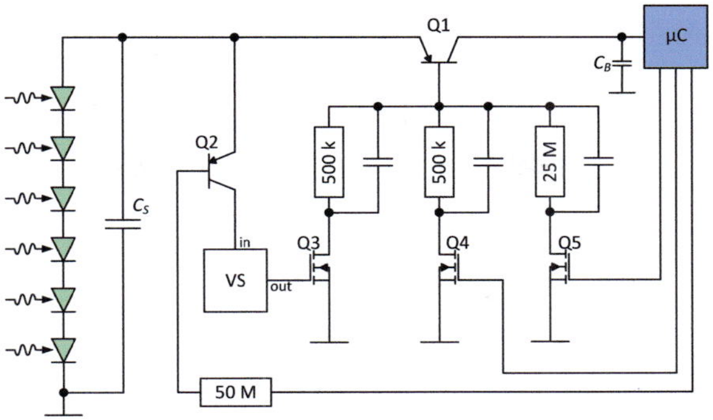

Fig. 4.28 Start-up electronic for the optically powered sensor node. The PNP transistor Q1 can be switched into conducting state by three independent FET (Q3 to Q5) with different resistors. Q1 connects the photodiode array and the energy storage capacitor C_S to the microcontroller (µC), Q2 to the voltage supervisor (VS). The microcontroller supply is additionally buffered with an buffer capacitor C_B.

Current Measurement of the Start-up Electronic

The current consumption in dependence of the applied voltage during the start-up sequence was measured at room temperature and is shown in Fig. 4.29 (a). The storage capacitor was not connected. The threshold of the voltage supervisor was set to its maximum value of 2.56 V. This assures that the microcontroller stays shut down during the measurement.

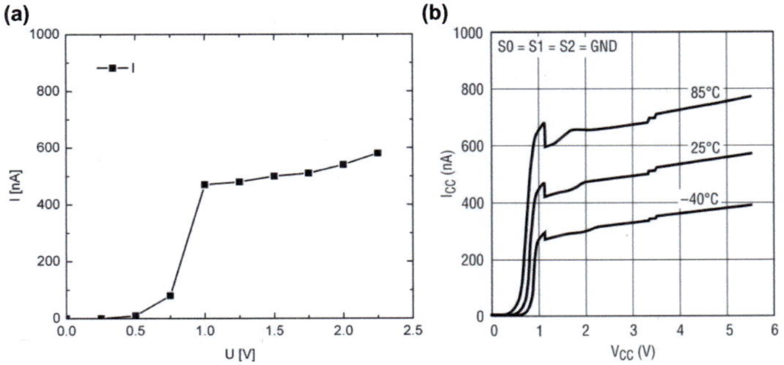

Fig. 4.29 Current consumption of (a) the whole start up electronic for increasing input voltage U at room temperature and (b) of the voltage supervisor, taken from the datasheet [62]. A comparison shows that the current consumption of the start-up electronics is dominated by the voltage supervisor. The three switches are only responsible for a small amount.

In Fig. 4.29 (b) the current consumption of the voltage supervisor is shown, taken from [62]. A comparison shows that in the start-up sequence the total current consumption is dominated by the voltage supervisor. Above 0.7 V the current increases rapidly to about 500 nA same as the voltage supervisor. The switches connected to the supply channel do not contribute significantly. The 1.9 V threshold is reached after 34 minutes with a total optical input power of $-18\,\text{dBm}$ (15.8 μW) to the sensor unit.

Microcontroller [36]

For energy saving the microcontroller can be operated in different low power modes (LPM). With increasing level of the LPM more and more functionalities are disabled. In LPM3 only an auxiliary clock is maintained which allows the microcontroller to reactivate itself after a programmed time. The auxiliary clock can be sourced by two different oscillators: an internal voltage controlled oscillator ("very-low-power low-frequency oscillator, VLO) and an external quartz oscillator. With the VLO the current consumption is only $I_{\mu C,LPM3} = 300\,\text{nA}$ but the frequency drift can be up to several percent (0.5% per K). With the external quartz 20 ppm accuracy is feasible but current consumption increases to 900 nA. Reactivation from LPM3 into the active mode takes less than 1 μs.

Energy Storage

For calculating the total power budget the internal losses in the storage capacitor have to be considered. This storage is the largest capacitor in the sensor unit and leakage currents of electrolyte capacitor are proportional to its capacitance [88]. However leakage currents are of often not specified precisely, therefore the leakage current of the nominal low-leakage capacitor was measured. The capacitor with capacitance C under test was initially charged to $U_0 = 10\,\text{V}$ and the capacitor voltage was measured over time. Internal leakage mechanisms let an internal current flow which results in loss of charge and voltage at the capacitor over time.

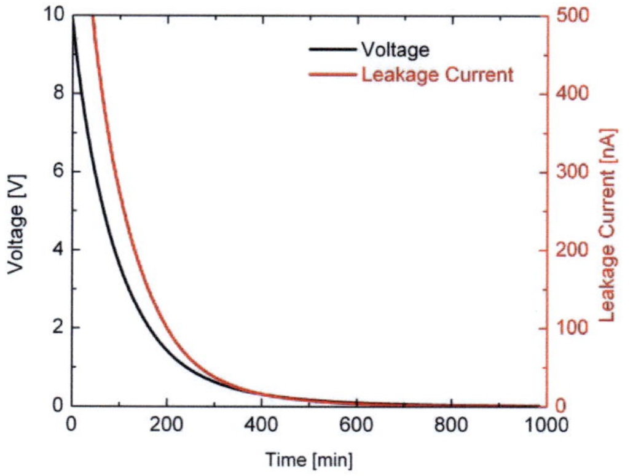

Fig. 4.30 Leakage of an electrolyte capacitor (left axis) Measured voltage over time, depicted in black, and (right axis) calculated leakage current thereof, depicted in red, of a charged capacitor [88]. The capacitor was charged with 10 V and the capacitor voltage was measured over time. Internal leakage mechanisms let an internal current flow which results in loss of charge and voltage at the capacitor.

Voltage measurements were performed with a voltmeter showing an input impedance of $>10\,\mathrm{G\Omega}$. The voltage drop over time is shown Fig. 4.30 in black. As expected the voltage drops exponentially over time,

$$U(t) = U_0 e^{-\frac{t}{RC}}. \tag{4.1.7}$$

The internal leakage mechanism is modeled here as a resistance R which is in parallel to the capacitor. From the measured voltage over time the resistance can be calculated to $13.9\,\mathrm{M\Omega}$. The measurement error coming from the current flow through the voltmeter can be neglected as the internal resistance of the capacitor is about a factor of 1000 smaller than the one of the voltmeter. Also the internal leakage current I_{leak} can be calculated from the voltage drop over time,

$$I_{\mathrm{leak}}(t) = \frac{\partial Q}{\partial t} = C\frac{\partial U(t)}{\partial t}. \tag{4.1.8}$$

In Fig. 4.30 the calculated leakage current is depicted in red. The maximum charge voltage is expected to be 2 V. For this voltage the leakage current can be read out to be 150 nA.

An upper bound of the total current consumption of the sensor node in its energy saving LPM3 can be estimated by the sum of microcontroller supply current $I_{\mathrm{\mu C,LPM3}}$, switch current I_{switch}, and leakage current I_{leak} in the storage capacitor,

$$\begin{aligned} I_{\mathrm{LPM3}} &= I_{\mathrm{\mu C,LPM3}} + I_{\mathrm{switch}} + I_{\mathrm{leak}} \\ &= 300\ \mathrm{nA} + 52\ \mathrm{nA} + 150\ \mathrm{nA} \\ &= 502\ \mathrm{nA}. \end{aligned} \tag{4.1.9}$$

At a total optical input power of –18 dBm to the sensor node, the diode array delivers a maximum current of 1.4 μA. This shows that in LPM3 the surplus of current charges the capacitor which than can be partially discharged in an active phase.

Total Power Budget

In different operation modes of the sensor node, different subsystems are powered up (like DC/DC boosters, receiver, transmitter, or attached sensors), or they run in different operation modes (like the microcontroller). For the different operation modes the respective current consumption measured at 2 V input voltage are listed in Table 4.2.

Table 4.2 Current consumption of the sensor node in different operation modes measured at an input voltage of 2 V

Operation Mode	Current Consumption
μC clocked by VLO (LPM3, Sleep mode)	0.55 μA
μC clocked by external quartz (Snooze mode)	1.00 μA
μC active (8 MHz)	1.58 mA
μC active (8 MHz) + DC/DC booster (no load)	1.88 mA
μC active (8 MHz) + DC/DC booster + Rx	11.10 mA
μC active (8 MHz) + DC/DC booster + Tx (operated in CW)	11.26 mA

Attachable Sensors

The chosen microcontroller can handle different digital interfaces and also offers an analog input. In order to design a flexible system the sensors were not directly soldered to the node. Sensors can be connected via pin headers to the digital or analog input of the microcontroller. Beside the data ports, pins for power supply are also provided. To prove the flexibility of the system three sensor submounts equipped with different sensors were designed. Different interfaces as well as different measurements were realized and are listed in Table 4.3.

Table 4.3 Different sensors attached via different interfaces to the sensor node.

Sensor type	Interface	Current Consumption	Manufacturer	Product Identifier
Temperature	SPI	0.5 mA	Maxim	DS1722 [89]
Acceleration	I^2C	2.3 mA	Memsic	MXC6202xJ/K [90]
Light	Analog	2.5 mA	Avago	APDS-9003 [91]

Demonstrator

The full electronic circuits (schematics as well as layouts) can be found in Appendix A.4. Three demonstrators were set up and a photography of one of them can be seen in Fig. 4.31 together with the sensor submounts. Fig. 4.31 (a) shows the sensor node. Via the connector at top left, the node is attached to the network. The optical input signal is split up at the optical splitter (bottom) and distributed equally to seven photodiodes. Photodiodes and splitter are connected with the couplers on the left hand side of the board. On the eighth port of the splitter the laser is connected. In the middle of the aluminum plate the electronics of the sensor node is hosted.

(a) (b)

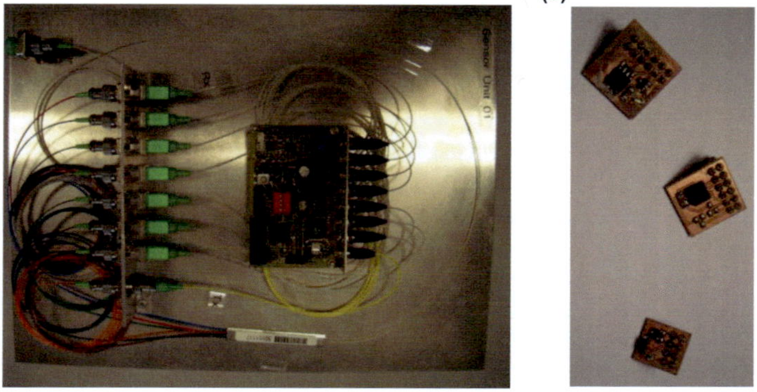

Fig. 4.31 Photograph of the (a) optically powered sensor node and (b) the attachable sensor submounts.

Fig. 4.31 (b) shows three attachable submounts. From top to bottom the boards carrying an SPI temperature sensor [89], an I^2C acceleration sensor [90], and an analog light sensor [91].

4.2.3 Applications with Low Power and Low Bandwidth

Sensor networks are increasingly deployed for home automation, facility management, monitoring the structural integrity of buildings and bridges, and high-voltage applications. As a power supply is needed for nearly all kind of sensors, unwired sensors need an inbuilt energy source like a battery, or have to harvest their energy from their environment. Thus, a limitation of lifetime or application area results. Instead of supplying power by electric wiring, our sensor nodes are supplied by optical power delivered in a fiber, which is required anyway for the data exchange with the base station. This enables operation with negligible susceptibility to electromagnetic interference and lightning due to the galvanic isolation between sensor nodes and base station. Likewise, operation in a discharge-sensitive environment and operation without electromagnetic radiation from wires even at high and highest data rates become feasible. Furthermore, the low loss of optical fibers enables to bridge large distances between sensors and controlling base station. Emphasis is on low power consumption for this kind of sensor nodes to achieve long reach, to meet security aspects, and to lower costs.

A recently emerging and very promising application field for such optically powered sensor networks is the envisaged smart power grid. As generation and consumption of electrical power will strongly vary in time, storage and routing of electrical energy will be mandatory. Therefore, measuring of the available and needed power as well as monitoring of the components of the power grid will strongly gain importance [92]. Small optically powered sensor nodes are promising candidates for simplifying this task by offering sufficient bandwidth, long reach, and perfect galvanic isolation. Future monitoring devices combine low- and high-latency services which can be handled favorably by the LE-MAC protocol. Another application of optically powered devices can be found in telecommunications. At so-called demarcation points, a traffic-transparent energy-autarkic sensor (demarcation device) is placed. It can be interrogated by the central office to monitor the link performance independently of an attached ONT [93].

5 Applications with High Power and High Bandwidth Demand

In the following sections, results of an optically powered video link are discussed. The results in Section 5.1 have been published in 2008 [J5]. Further developments in electronics and optics led to improvements which were studied in the Diploma Thesis of Michael Dreschmann in 2009 [S10] and published as conference contribution in 2010 [C2]. These results are presented in Section 5.2.

5.1 Video Transmission System without Control Channel

An Optically Powered Video Camera Link

G. Böttger, M. Dreschmann, C. Klamouris, M. Hübner, M. Röger, A.W. Bett, T. Kueng, J. Becker, W. Freude, and J. Leuthold

in *IEEE Photonics Technology Letters*, vol. 20, pp. 39–41, January 1, 2008.

An optically powered camera sensor link is demonstrated. Power and data are transmitted over a 62.5-μm multimode glass fiber. Uncompressed video with 640×480 pixels resolution is streamed continuously at 100 Mb/s as soon as the fiber is illuminated with sufficient optical power. No energy has to be stored at the sensor location in batteries with limited capacities and lifetimes. Inexpensive fiber optics and low-power state-of-the-art electronics are used to make > 100 mW available at sites which have no direct access to an electrical network. There is a complete electrical isolation between the remote camera unit and a base station.

A landmark paper in the late 1970s suggested and demonstrated optically powered speech communication [45]. In this scheme, the power to drive remote sensors or communication units is transmitted along with data, ideally over a single optical fiber. An inherent advantage of optically powered systems is the small susceptibility to electromagnetic noise and lightning, and their suitability in discharge-sensitive environments.

Early optically powered sensor systems were operated with lowest speed analog or digital optical signals in order to cope with the few milliwatts of available optical power [45] – [47]. For instance, high-voltage installations have been monitored by means of an optical signal [47]. In other implementations microelectromechanical systems were activated optically to perform remote switching [48], or splitting ratios of couplers in optical communication networks were adjusted [49]. A unidirectional transmission of data at 22.5 Mb/s was demonstrated, powered optically by a compact disc laser diode (LD) [50], and also local area network applications have been tested [51], [52]. More recently, photovoltaic (PV) converters in series have been used to provide higher voltage swings, and thus allow more complex electronic circuits [53], [54]. However, so far, due to power limitations, demonstrations were restricted to very simple functions. Demanding applications that require sampling, generation, and transmission of moderate digital bit streams have not been realized. In this section, we report online remote optically powered video sensing and streaming with a video graphics array resolution at 100 Mb/s. Using an optimized single-cell PV converter, we drive state-of-the-art electronics with voltages around 0.80 V. A serialization of converter cells is not required, keeping the converter costs low. Full-color video transmission at 15 frames per second is demonstrated. The sensor platform is easily configurable to host a multitude of other physical sensors.

System Configuration

A schematic of the optically powered video camera is shown in Fig. 5.1. The base station and remote camera unit are connected with a single gradient-index multimode fiber (GI-MMF, 62.5-μm single core by OFS-Optics). All power required for the operation of the remote unit is supplied by a high-power LD at the base station [see Fig. 5.1 (left)]. The high-power LD provides up to 1-W fiber-coupled optical power at 810-nm wavelength, for which sources have become very reliable and inexpensive. The uplink data channel for transmitting a video stream from the optically driven remote unit transmits at 1310 nm. Power and data channels are combined into the single fiber connecting base station and remote unit by respective 810 nm/1310 nm thin-film-filter-based wavelength-division couplers. A downlink channel for controlling the remote unit does not exist.

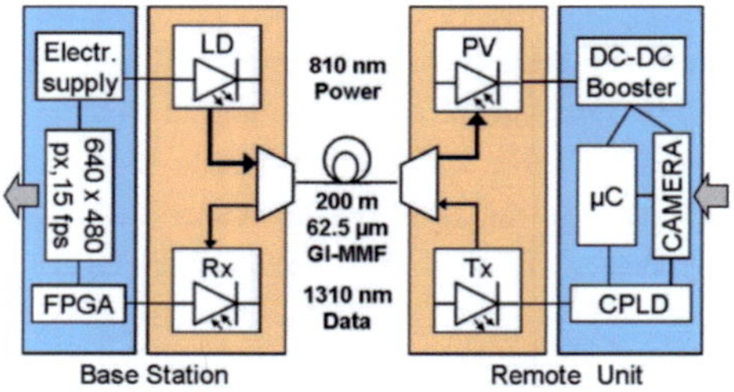

Fig. 5.1 Schematic of data and power transmission over fiber. An 810-nm wavelength LD in the base station drives remote unit electronics via a PV converter cell. Data acquired on the remote unit are encoded on a 1310-nm LD (Tx), and coupled into the same fiber supplying the power. Received video data are decoded in the base station with an FPGA. Data are transmitted in one direction only. Other abbreviations, see text.

A key component of the remote unit is a fiber-illuminated PV converter [Fig. 5.1 (right)]. Details of the opto-electrical conversion are discussed in the following section. The converter cell output voltage is raised and stabilized with an electrical–electrical voltage converter (direct current DC–DC booster). This way complex electronics assembled on the remote unit can be operated at a standard 2.5-V voltage level. Electronics in the remote unit consist of a low-power complementary metal-oxide semiconductor (CMOS) camera sensor, a complex programmable logic device (CPLD) for serializing 8-bit-parallel camera data, and a 16-bit general purpose microcontroller (μC) storing initialization routines. Data acquired and processed on the remote unit are directly modulated at 100 Mb/s onto the 1310-nm data channel LD. Details on data acquisition, transmission, and decoding are given in a later section.

The optical attenuation in the fiber channel is 2.3 dB for an exemplary 200-m fiber link between base station and remote unit. This number includes the 810/1310-nm wavelength-multiplexing coupler losses that contribute 0.6 dB each, fiber attenuation at 810 nm in the order of less than 0.8 dB for 200-m distance, fiber splices and connector losses that account for another 0.3 dB. For camera link operation the remote unit circuits require an electrical power of 103 mW at 2.5 V. The electrical–electrical voltage conversion from 0.8 V of the PV cell is > 75%. With this load attached, the actual optical–electrical PV conversion takes place with 33 % efficiency. Factoring in all transmission and conversion losses, a total of 670 mW fiber-coupled power needs to be launched in the base station.

PV Operation of Remote Unit

All electrical energy required for the remote unit sensor circuitry is supplied by a miniaturized PV cell as depicted and characterized in Fig. 5.2. The single-cell GaAs-based PV converter is optimized for illumination with 810-nm wavelength, and supplies an open circuit voltage of $U_{oc} = 1.15\,\text{V}$. Fig. 5.2 (b) shows an enlarged view of the active area of the cell (1 mm diameter). Width and height of the top electrode grid fingers were optimized to minimize shading losses while efficiently collecting photo currents. The single-cell design is tolerant to partial illumination, unlike converter configurations serializing multiple cells. Keeping a single-cell layout also reduces the number of processing and connectorization steps. The circular cell design is well matched to illumination from fibers, its active area may be exposed to optical power densities up to 50 W/cm. This is typically 500 times higher than in conventional, large-scale PVs. Opto-electrical conversion maximally is 50 % for a power density of 6.5 W/cm² (open circuit case, no load attached). Comparable power conversion efficiencies in similar material systems have been reported in the past, the record being 56 % with an external quantum efficiency > 90 % [63]. The PV cell used in this setup was designed and processed by Fraunhofer ISE (Institute for Solar Energy Systems). For details on the metal–organic vapor-phase epitaxy layer structure, refer to [64].

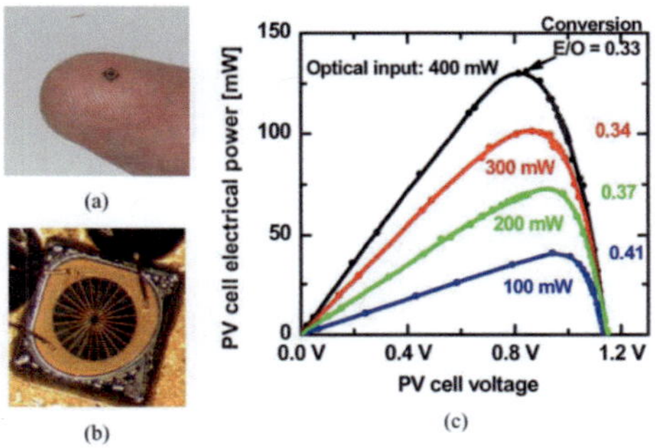

Fig. 5.2 (a) Unmounted single-cell PV converter with a diameter of 1 mm; (b) enlarged view of cell mounted on a metal casing. (c) Remotely available converted electrical power for four optical illumination levels. O/E conversion efficiencies given for maximum power extraction.

The electrical power as a function of the PV cell voltage for different optical input powers are plotted in Fig. 5.2 (c). The curves are obtained by sweeping an external load resistance from short circuit $(R_L \rightarrow 0\,\Omega)$ to open circuit $(R_L \rightarrow \infty)$. The plot shows that for an illumination with 400 mW of optical power (50 W/cm), a total of up to 133 mW electrical power can be harvested. To run the device at its optimum conversion efficiency, it needs to be operated at a load resistance around $R_L \approx 5\,\Omega$. The PV cell voltage then is 0.8 V, which is sufficient to start up the oscillator-stage of the electrical–electrical converter (DC–DC booster). The respective opto-electrical (O/E) conversion efficiency at this best operation point then becomes 33 %. Higher O/E conversion efficiencies can be attained for smaller illumination levels. With reduced illumination, there is less internal thermal heating and in consequence inner resistance decreases [64].

Video Data Acquisition and Transmission

At the location of the remote unit, a full-color video signal is acquired and digitized by a low-power CMOS camera chip. The continuous video stream has a resolution of 15 frames per second and 640×480 pixels (8-bit uncompressed red/green/blue data). The camera chip consumes 40 mW at a clock cycle of 12.5 MHz. A low-power microcontroller (Texas Instruments MSP430 series [36]) initializes the camera chip over an I^2C interface (intelligent interface controller). Video data from the eight digital output ports of the camera chip are then aggregated at 100 Mb/s and Manchester-encoded with an XOR stage in a complex programmable logic device (128 cell CPLD Xilinx Cool-Runner). The CPLD together with an external clock oscillator is estimated to dissipate another 40 mW. The encoded signal then directly modulates a 1310-nm wavelength LD. The LD is operated slightly above its low threshold current of 10 mA, and couples 2 mW of optical power into the fiber.

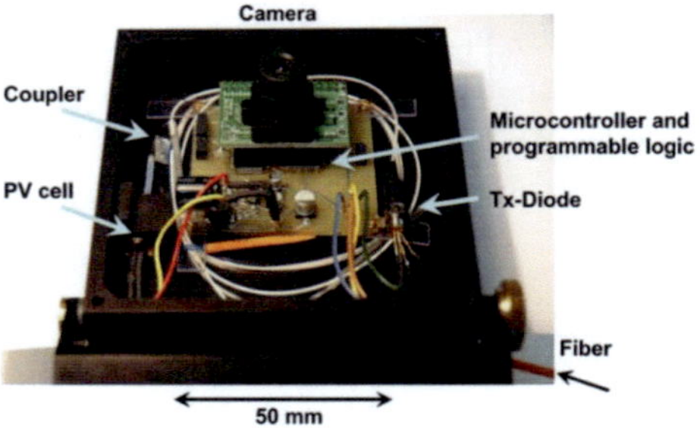

Fig. 5.3 Optically powered remote unit with connecting fiber at bottom right. All electronics in the unit are driven after O/E conversion by the high-efficiency PV converter (PV cell).

A wavelength of 1310 nm was chosen for the data transmission because at > 100-nm distance from the 810-nm power channel it can be easily separated with inexpensive, thin-filter-based wavelength-division couplers. Attenuation for the used MMF at 1310 nm is merely $\approx 0.5\,\mathrm{dB/km}$. Transmitting data at 100 Mb/s over 200 m uses a fraction of the specified $580\,\mathrm{MHz}\times\mathrm{km}$ fiber-bandwidth-length product, but requires using GI fiber instead of step-index fiber [51]. We did not observe detrimental effects from guiding power and data channels in the same fiber. Given the 62.5-μm core, power densities remain moderate, and are uncritical even for fiber connectors. An optically powered remote unit prototype is shown in Fig. 5.3. The PV cell for O/E power conversion is mounted on a metal bracket for thermal dissipation.

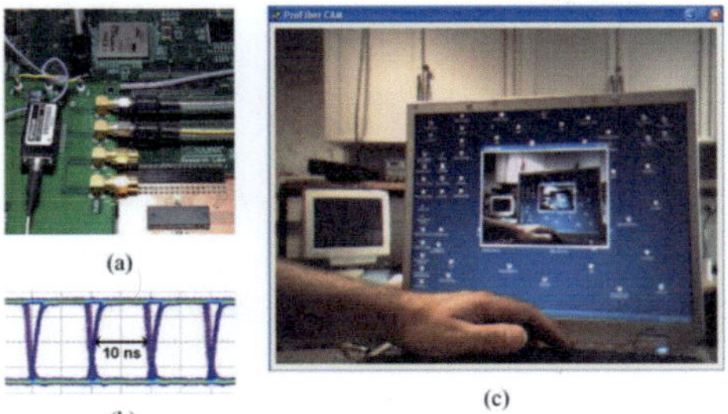

Fig. 5.4 (a) Base station receiver diode connected to IO-ports of an FPGA-board sampling and decoding video data. (b) Eye diagram of decoded data after 200 m fiber length. (c) Optically powered video transmission viewed in a 640×480 pixel window on a laptop computer display.

After transmission through the optical fiber link, data are detected at the base station with a standard receiver diode [see Fig. 5.4 (a)]. The dynamic input range of the receiver reaches from –3 to –38 dBm. With a data signal power of approximately –3 dBm in the base station before the receiver there is ample system margin. Thin-film filters in the couplers easily realize an isolation of > 30 dB between power and data channels, required for avoiding crosstalk into the data channel [52]. Bit-error-rate tests were conducted and showed error-free performance. A clear and open eye diagram as received under the above conditions is shown in Fig. 5.4 (b). Synchronization of the video stream is guaranteed by sending and detecting control frames. For clock extraction, the 100-Mb/s data are sampled with 2 Gbit/s on differential-input ports of a field-programmable gate-array (FPGA) board (Xilinx Spartan 2). The Manchester signal is decoded on the board, video information color-converted and buffered in graphics memory for viewing on displays as in Fig. 5.4 (c).

Conclusion

We have demonstrated an optically powered video camera link that allows acquiring and communicating a 100-Mb/s video stream over hundreds of meters distance. The remote unit and base station are connected by a single multimode glass fiber that is immune to electromagnetic noise or power surges. Effectively, 133 mW electrical power were made available on the remote unit by means of a miniaturized PV converter cell. It was shown that complex, programmable sensor circuitry may be operated. At present, a point-to-point connection has been realized. Any common passive optical network topology, however, can be implemented. Identifying sensor nodes in such networks would be simple, because unique addresses can be stored in electronic memory. This is less costly than, e.g., using fiber Bragg-gratings in all-optical sensor networks. The general sensor platform is well suited for sensor applications in hazardous and electrically sensitive environments. The used components are inexpensive, with further potential in cost and size reduction. Transmitting power and data over a single-core MMF proved to be uncritical, permitting high-data-rate light-weight installations. With ever-evolving electronics, we expect to see an increase of processing power and a further growth of sensor complexity, which makes the "power-over-fiber" scheme even more attractive.

5.2 Video Transmission System with Control Channel

Reconfigurable hardware for power-over-fiber applications

M. Dreschmann, M. Hübner, M. Röger, O. Sander, C. Klamouris, J. Becker, W. Freude, and J. Leuthold

in *Proc. 20th International Conference on Field Programmable Logic and Applications (FPL)*, Milano, Italy, August 2010. Reprinted, with permission, from [C2] © 2010 IEEE

The above described system has been developed further and the results of this development were published in 2010 [C2]. Most important improvements are a new photovoltaic converter delivering up to 2 V under high power illumination, the installation of a downlink data channel from base station to remote unit, an increase in the uplink data rate from remote unit to base station to 160 Mbit/s, and the equipment of the remote unit with servo motors to manipulate the camera module. Steering of the camera module is done via a web interface hosted in the base station, which sends the manipulation commands via the optical downlink communication channel to the remote unit.

System Overview

The optically powered remote platform is connected to the base station by an optical fiber which carries power in form of light and control data from the base station to the sensor (downlink), and by a second fiber for the transmission of video and further telemetric data back to the base station (uplink) as illustrated in Fig. 5.5. With the help of wavelength splitters it is also possible to carry both optical signals over one fiber as we demonstrated in [65].

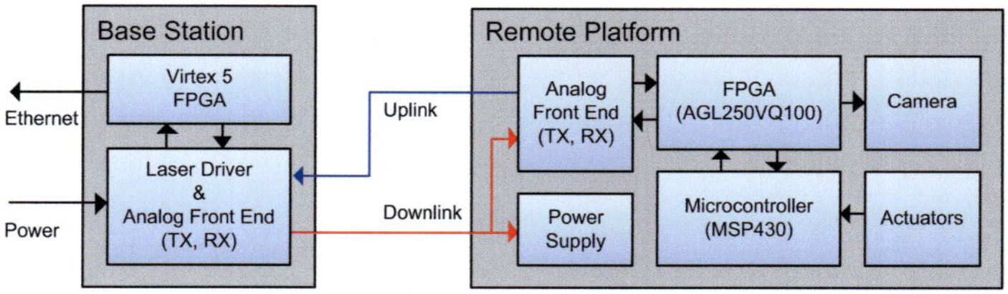

Fig. 5.5 System overview

The remote platform shown in Fig. 5.6 consists of two printed circuit boards (PCB) connected by high density connectors. The lower PCB holds the optical interfaces and the signal processing components, namely an MSP430 microcontroller and an Actel Igloo field-programmable gate array (FPGA). The upper PCB serves as the actual platform for the complementary metal-oxide semiconductor (CMOS) video sensor and the motors. So the sensor and actuator types can be changed easily for tailoring the sensor platform to different applications. Due to the fine granular reconfigurability of the Actel Igloo FPGA a wide range of types of sensors and actuators may be connected directly omitting the need for power consuming interface circuits.

Fig. 5.6 Optically powered sensor platform

The base station transmits energy and control data to the remote platform and receives video and telemetric data from it. The signal processing part of the base station is based on a Xilinx Virtex-5 FPGA which holds a LEON-3 system-on-chip from Gaisler Research [66]. The digital hardware of the remote platform and the base station need appropriate interfaces to the glass fibers connecting both platforms. These analog frontends are described in the next section.

Analog Front-Ends

Fig. 5.7 shows a simplified version of the interface circuit used to connect the downlink glass fiber to the base station's FPGA. Basically the circuit provides an adjustable current source which drives a high power laser diode (HP-laser diode) whose optical output power is transmitted to the remote platform via the glass fiber. Additionally the laser diode current and therefore its output intensity can be modulated by a data signal which is generated by the FPGA.

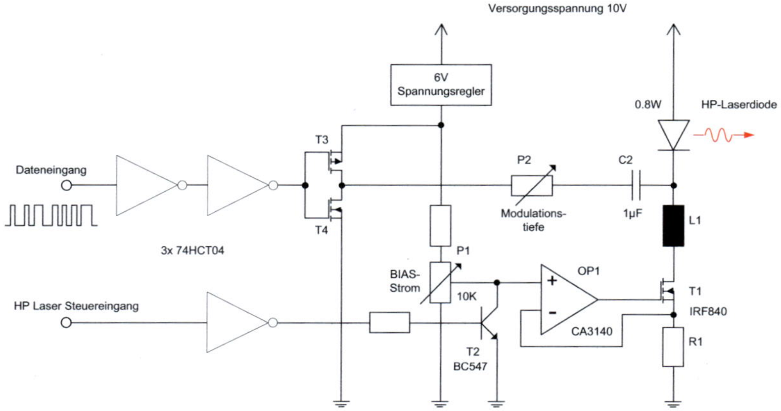

Fig. 5.7 Laser driver at the base station

To provide the necessary power to the remote sensor, a laser diode with an optical output power of up to 1.5 W is used. However, for a testbed with 200 m of fiber between base station and sensor we only needed 800 mW of optical output power to provide 320 mW of electrical power to the sensor platform. For an optical output of 800 mW, the laser diode needs a current of approximately 1.8 A which is regulated by operational amplifier OP1 in conjunction with transistor T1 and shunt resistor R1. The digital control data signal is amplified by the transistors T3 and T4 and then coupled capacitively into the main laser diode current by C2. Due to the capacitive coupling a DC-free line code like Manchester needs to be used. The 74HCT04 inverters accomplish the level shift from the FPGA outputs operating at 3.3 V to a voltage of 6 V used by the rest of the modulation circuit.

At the remote platform the signal of the downlink fiber is received by a photovoltaic converter (PVC) and needs to be separated back into its power and its data component. This is done by an LC filter circuit as illustrated in Fig. 5.8.

The power component is isolated by inductor L1 and charges a 50 F gold-cap capacitor. Depending on the actual load, this primary input voltage of the remote module is in the range between 1.8 and 2.2 V. The gold-cap capacitor acts as an energy reservoir to supply the actuators with up to two watts of power upon request. The data signal which is modulated on the power signal of the downlink is extracted by capacitor C2 and amplified by a subsequent transimpedance amplifier and a comparator. Its output is subsequently processed by the Actel Igloo FPGA located at the remote platform.

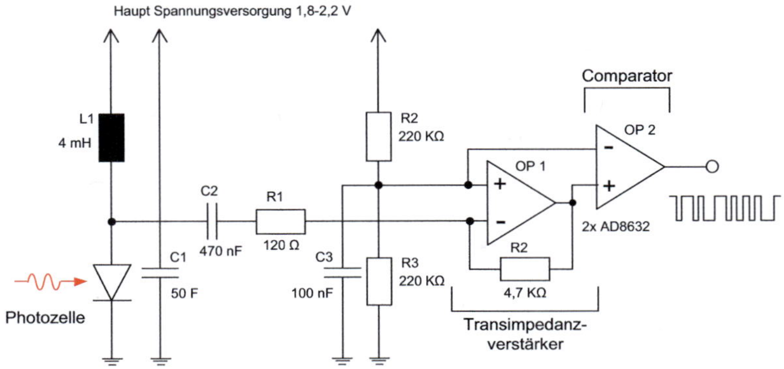

Fig. 5.8 Optically powered sensor platform – receiver frontend

Experiments showed that a square wave signal of up to 200 kHz may be transmitted via the downlink. The limitation in speed is due to restrictions of the PVC and the high power laser diode at the base station. If a higher data rate is required for the downlink, this can be achieved by a separate optical link. However, the control data carried by the downlink is only required to transmit desired positions for the actuators and some configuration options for the camera sensor and hence the bandwidth still is sufficient. To further increase the transmission reliability a Manchester encoded signal of only 100 kbit/s is used.

The contrary is true for the uplink where the uncompressed video stream data from the camera sensor need to be transmitted. For that purpose at least 128 Mbit/s are necessary. Data is transmitted with an 8B10B line coding so that the resulting bit rate equals to 160 Mbit/s. Because the uplink does not need to transmit power, a standard optical transmitter and receiver pair suitable for the desired data rate could be used.

The primary input voltage of approximately 2 V is only appropriate to power the MSP430 of the remote platform. The FPGA and further components like the camera sensor and the actuators need other voltage levels and better voltage stability. These voltages are generated by a set of switching mode power converters. For the FPGA and the camera sensor they provide a stabilized output of 1.5 and 2.5 V, respectively. A 5 V output to power the actuators is generated upon request. Because costs of the optical components strongly depend on the amount of optically transferred power, high efficiency of the power supply circuits at the remote platform is very important. The used converters from TI (TPS61030 [67] and TPS62000 [68]) provide a very high efficiency of up to 96 %.

Remote Platform

Besides the analog front end of the remote platform, an Actel Igloo FPGA (AGL250VQ100, [69]) and a MSP430 microcontroller from TI (MSP430F149, [36]) are required to handle the signal processing at the optically powered remote platform. They are connected to the actual camera sensor (OV7640 from Omnivision [70]) and the actuators which are two small servo motors.

The microcontroller is responsible for the power management on the remote sensor platform and controls the switching mode power converters and hence the power distribution to the other parts of the sensor. Furthermore its integrated analog-digital converter is used to monitor all power supply voltage levels and its pulse width modulation channels control the movement of the servo motors used to change the position of the camera sensor.

The main responsibility of the FPGA located in the remote platform is receiving the video data from the camera sensor and transmitting them to the base station. The FPGA also decodes the incoming control data from the base station, which is received by the analog front end. The output of the CMOS sensor is a 4:2:2 chroma subsampled YCbCr signal with 8 bit resolution per component which is directly sent to the base station. The FPGA is controlled by the microcontroller over a serial peripheral interface (SPI) connection and allows the controller to receive the control data and insert command responses and telemetric data into the uplink data stream. The relationship between microcontroller and FPGA is illustrated in Fig. 5.9.

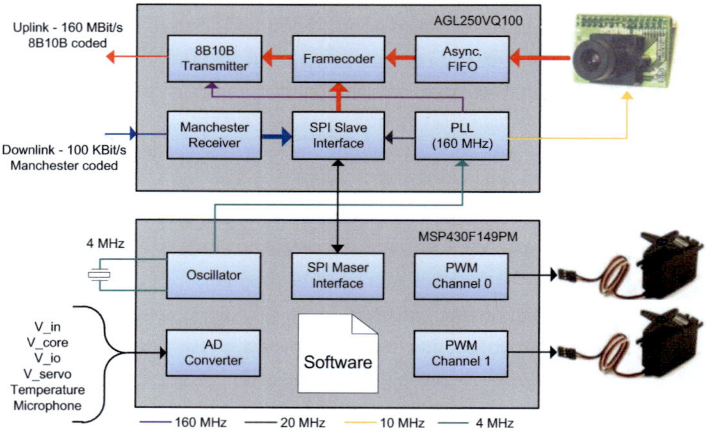

Fig. 5.9 Signal processing of the remote sensor platform

The data transfer between the base station and the remote platform is packet based. The packet structure is identical for downlink and uplink. A preamble of four bytes is followed by a length code of two bytes, a two byte type code, up to 65534 data bytes and a 32 bit cyclic redundancy check (CRC). The distinction between different types of data is done by the type code as listed in Table 5.1.

Table 5.1 Packet types

Packet Type	Description
0x0000 – 0x01DF	Video Data
0x01E0	End-of-Frame Marker

Packet Type	Description
0x1000	Audio Data
0xF000	Command Data
0xFFFF	Filler Packet

As one can see, there are 480 different packet types for video data. Each type corresponds to one line of the video frame which has a resolution of 640×480 pixels. The data of 640 pixels from one line is transmitted in one packet. Because of the 4:2:2 chroma subsampling, one pixel occupies two bytes and therefore 1280 data bytes are required per line and video packet.

The frame coder shown in Fig. 5.9 is responsible for the formation of the miscellaneous data packets from Table 5.1. Besides the video data also command data and audio data coming from the MSP430 via SPI is processed. If no data is ready to be transferred, the frame coder inserts filler packets. They contain dummy data and thus ensure that the uplink does not get idle. This is necessary, because the analog receiver circuit at the base station needs some microseconds to recover its optimal operation conditions after a longer idle time thus resulting in data corruption.

Base Station

The signal processing part of the base station is based on a Xilinx Virtex-5 FPGA (XC5VLX110T, [71]) which holds a LEON-3 system-on-chip (SoC) from Gaisler Research [66]. The SoC is mainly build of the LEON-3 CPU core, a DDR2 memory controller (to access 256 MB of external RAM provided by the ML509 board), an Ethernet controller, a VGA controller and an UART from the LEON-3 library (see Fig. 5.10). Two additional modules, a module for the communication with the sensor platform and a hardware JPEG encoder where implemented newly.

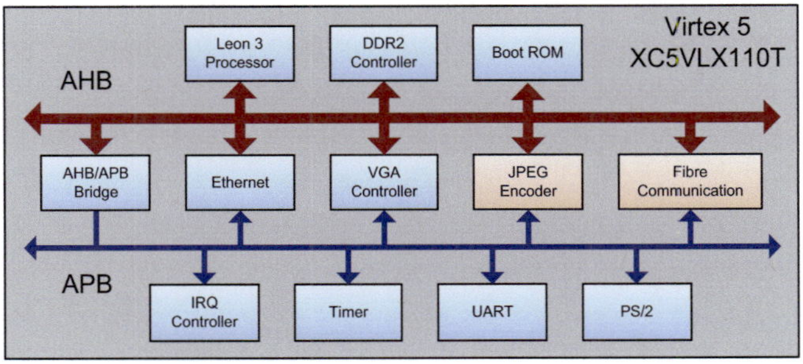

Fig. 5.10 SoC overview of the base station

Evaluation

Table 5.2 shows the power consumption of the individual components of the remote platform. The Actel Igloo FPGA only consumes 80 mW. When the servo motors are inactive, the whole sensor platform requires 160 mW of electrical power. Because in total 320 mW arrive at the platform, another 160 mW are available to charge the gold-cap capacitor.

Table 5.2 Power consumption of the remote platform

Component	Power consumption
MSP430	1 mW
Analog Front End (RX)	0.5 mW
Actel Igloo FPGA	80 mW
CMOS Sensor	50 mW
Analog Front End (TX)	20 mW
Servo motors (max)	1800 mW
Total (with motors inactive)	160 mW

Some typical performance data of the demonstrator is illustrated in Table 5.3. The actual video frame rate is limited to 12.5 frames/s due to a layout error at the sensor PCB. However, the general capability of the demonstrator to process up to 25 frames/s was verified by a test image generator implemented in the Actel Igloo FPGA.

Table 5.3 Performance of different components and data links

Component / Data Link	Performance	
Downlink	100	kbit/s
Uplink (effective)	128	Mbit/s
Uplink (line)	160	Mbit/s
Video frame rate (actual)	12.5	frames/s
Video frame rate (max)	12.5	frames/s
LEON-3 SoC operating frequency	120	MHz
JPEG Encoder (@ 120 MHz, 640x480)	12.5	frames/s

Table 5.4 shows the logic occupation of the utilized FPGAs. Both have still enough free resources for further enhancements.

Table 5.4 Logic occupation of the utilized FPGAs in the remote platform (Actel) and in the base station (Virtex 5)

Component	Logic Occupation	
Actel Igloo FPGA	2,539 Core Cells	(41 %)
Virtex 5 FPGA	15,713 V5-Slices	(22 %)

Conclusion and Future Work

Our system demonstrates that it is possible to assemble an optical powered video camera system which is as powerful as ordinary systems but also comprises the typical advantages of pure optical powered systems like an advanced robustness to electromagnetic interferences. The optically transferred power is not only sufficient to operate a CMOS video sensor and transmission logic but also allows us to drive actuators which are able to move the sensor.

The experiment further shows, that parallel working, FPGA based hardware has advantages over a traditional approach with low power processors. Additionally, the hardware can be used to reduce the communication effort by including preprocessing functionality. This could be e.g. filters for image processing or pattern recognition algorithms. With this functionality the remote sensor could detect a predefined scenario in the pictures and send a message to the base station which could be useful for surveillance cameras. Due to the fact that reconfigurable hardware can be re-used with other algorithms, the functionality on the sensor can be reconfigured if required. This flexibility enables a high degree of freedom for the camera approach which then can be deployed in different applications.

While there is a wide potential to improve the presented sensor platform as well as the base station in terms of power consumption, performance and usability, future work will also include the design of more complex optical powered networks with a wider spectrum of sensor types as mentioned in [72]. Such an advanced network would include several sensors which are operated simultaneously in the network as well as optical splitters and switches to optically connect the different sensors to a base station. The splitters and switches allow the distribution of power and communication bandwidth to the sensors depending on their actual needs. This requires also new communication protocols and advanced power management techniques at the involved remote sensor platforms and the base station.

Summary and Future Work

Optically powered sensor networks are a promising solution for sensor networks where standard electrical wiring or wireless techniques fail.

In this thesis, optically powered sensor networks have been investigated. Advantageous application fields have been identified and appropriate prototypes of different sensor nodes were developed, set up and characterized. Broad spans of power consumptions and data rates of the respective nodes were covered. Base stations to gather and display the information collected by the nodes and to supply the sensor nodes have been set up, too. Further, a special protocol taking into account passive optical fiber networks characteristics as well as different needs of different nodes has been developed. Improvements in power consumption of the nodes enabled through the special communication protocol were exploited.

Studies on Optical Power Supplying: For each node, the optical power supply was adapted to guarantee optimum performance. Different photovoltaic power converters as well as lasers were studied. Further, the circuiting of several converters was studied to optimize the power conversion efficiency especially for small optical powers.

In future, multi-segment photovoltaic converters optimized for small illumination power would allow for a more compact design of the sensor node. Decreases of the saturation current would lead to significant improvements in conversion efficiency. Even integration of a laser into the converter is possible. With all optical components integrated into a single housing inexpensive and compact sensor nodes are feasible.

Sensor Node and Base Station Design: Different sensor nodes were developed, set up and characterized. A node comprising a video module and servomotors for moving the module and nodes performing simple operations like measuring temperature, brightness and acceleration were presented. Appropriate base stations were set up also.

Electronics – of which the sensor nodes and base station are mainly set up – are a rapidly evolving field. Supply power of logic circuits and sensors go on reducing, whereas performance is increasing. So in future, sensor nodes may be operated with less power or will perform more complex operations. Together with more compact optical components the design becomes less complex and optically powered sensor networks might become commercially available in the near future. Even new application fields may be opened up where electronics sensors are not yet available.

This work was supported by the BMBF joint project “Components for Optical Monitoring of Access Networks (COMAN)”, funded by the German Ministry of Education and Research.

A. Appendix

A.1. Lower Limit for the Saturation Current of a Diode

The efficiency of photovoltaic conversion is influenced by numerous parameters. Many of them can be optimized by an appropriate choice of structure, material and operating condition. Nevertheless, some of these factors are limited by physical constraints, e.g., by the saturation current of a diode. The smaller the saturation current of a photodiode is, the larger its open-circuit voltage and therefore its conversion efficiency becomes, see Eq. (2.1.11). In the following section, the ultimate lower limit for the saturation current will be derived. It results from a detailed balance between emitting and absorbing photons in thermodynamic equilibrium [21]. The radiation of a black body in thermal equilibrium is described by Planck's law, which in the following will be applied to the radiative generation and recombination of carriers in a p-n junction [1].

Planck's law can easily be derived from photons (indexed with γ) captured in a cavity in vacuum with dimensions L_x, L_y and L_z and of volume V

$$L_x \cdot L_y \cdot L_z = V. \tag{A.1}$$

Since the photons are localized in the cavity, their momentum p of the states gets quantized in multiples of

$$\Delta p_x = \frac{h}{L_x}, \quad \Delta p_y = \frac{h}{L_y}, \quad \text{and} \quad \Delta p_z = \frac{h}{L_z}, \tag{A.2}$$

where h is Plank's constant. A single state occupies in momentum space a volume of

$$\Delta p_x \cdot \Delta p_y \cdot \Delta p_z = \frac{h^3}{V}. \tag{A.3}$$

Each state in momentum space belongs to 2 photon states coming from two orthogonal states of polarization. The spectral density of photons $n_\gamma(W)$ in the cavity in an energy interval $hf = W$ to $W + \mathrm{d}W$, where f is the frequency, depends further on the density of states $D_\gamma(W)$ and their distribution function $f_\gamma(W)$

$$n_\gamma(W) = D_\gamma(W) f_\gamma(W). \tag{A.4}$$

Photons have an integral spin and thus the Bose-Einstein distribution gives the average number of photons per mode of the electromagnetic field having energy $W = hf$

$$f_\gamma(W) = \frac{1}{\exp\left(\frac{W}{kT}\right) - 1}. \tag{A.5}$$

The quantity k is Boltzmann's constant, and T is the absolute temperature of the photon gas. To calculate the density of states $D_\gamma(W)$ we need the number of states $N_\gamma(W)$

$$D_\gamma(W) = \frac{1}{V} \frac{\mathrm{d}N_\gamma(W)}{\mathrm{d}W}. \tag{A.6}$$

From Eq. (A.3) we only know the momentum space a state occupies. This has to be transformed in energies. For photons, we find with the velocity of light c in vacuum

$$hf = c|p|. \tag{A.7}$$

So, the number of states $N_\gamma(W)$ is given by the spherical volume in momentum space with radius $|p|$ divided by the volume for one state given in Eq. (A.3) multiplied by 2 for the polarization state,

$$N_\gamma(W) = 2\frac{(4\pi/3)|p|^3}{h^3/V} = \frac{(8\pi/3)\cdot W^3}{c^3h^3/V}. \tag{A.8}$$

Inserting Eq. (A.6), Eq. (A.8) and Eq. (A.5) into Eq. (A.4) one finds the spectral density of photons (unit $1/\mathrm{m^3J}$)

$$n_\gamma(W) = \frac{8\pi W^2}{c^3h^3}\frac{1}{\exp\left(\frac{W}{kT}\right)-1}. \tag{A.9}$$

The spectral energy density $E_\gamma(W)$ (unit $\mathrm{J}/(\mathrm{m^3J})$) results by multiplying $n_\gamma(W)$ with the photon energy

$$E_\gamma(W) = \frac{8\pi W^3}{c^3h^3}\frac{1}{\exp\left(\frac{W}{kT}\right)-1}. \tag{A.10}$$

The black body radiates isotropically and the energy travels with the velocity of light c. The radidated spectral intensity $U_\gamma(W)$ (unit $\mathrm{W/m^2J}$) is then

$$U_\gamma(W) = E_\gamma(W)\frac{c}{4\pi} = \frac{2W^3}{c^2h^3}\frac{1}{\exp\left(\frac{W}{kT}\right)-1}. \tag{A.11}$$

The black body emits intensity $U_{\gamma,\mathrm{H}}(W)$ into halfspace $\Omega_\mathrm{H} = 2\pi$, where φ is the angle with respect to the normal of the emission area and ϕ is the azimuth angle,

$$\begin{aligned} U_{\gamma,\mathrm{H}}(W) &= \int_{\Omega_\mathrm{H}} U_\gamma(W)\,\mathrm{d}\Omega = \int_{\varphi=0}^{\pi/2}\int_{\phi}^{2\pi} U_\gamma(W)\sin\varphi\cos\varphi\,\mathrm{d}\varphi\mathrm{d}\phi = \pi U_\gamma(W) \\ &= \frac{2\pi W^3}{c^2h^3}\frac{1}{\exp\left(\frac{W}{kT}\right)-1}\mathrm{d}W. \end{aligned} \tag{A.12}$$

A diode in thermal equilibrium emits and absorbs the same number of photons from and to a halfspace which is at the same temperature as the diode [21]. Emission of photons happens by spontaneous recombination of electron hole pairs with rate r_0, absorption by generation of electron hole rate with rate g_0. In equilibrium, the rates are equal and are connected to the intrinsic carrier density n_i over the coefficient B for radiative ecombination,

$$g_0 = r_0 = Bn_i^2. \tag{A.13}$$

Only photons with energy $W > W_G$ are capable to generate electron hole pairs. For the generation rate follows

$$g_0 = \int_{W_G}^{\infty} \frac{U_{\gamma,\mathrm{H}}(W)}{W}\,\mathrm{d}W = \int_{W_G}^{\infty} \frac{2\pi W^2}{c^2 h^3} \frac{1}{\exp\left(\frac{W}{kT}\right)-1}\,\mathrm{d}W. \tag{A.14}$$

In equilibrium, the electron concentration n in the conduction band and the hole concentration p in the valence band are connected by the law of mass action (n_i is the intrinsic density),

$$np = n_i^2. \tag{A.15}$$

In non-equilibrium, the law of mass action is modified by an exponential factor

$$np = n_i^2 \exp\left(\frac{U}{U_T}\right) \tag{A.16}$$

with equivalent voltage U and thermal voltage $U_T = kT/e$, where e is the elementary charge. This modification also affects the recombination rate r in non-equilibrium

$$r = Bnp = \underbrace{Bn_i^2}_{=r_0=g_0} \exp\left(\frac{U}{U_T}\right) = \exp\left(\frac{U}{U_T}\right) \int_{W_G}^{\infty} \frac{2\pi W^2}{c^2 h^3} \frac{1}{\exp\left(\frac{W}{kT}\right)-1}\,\mathrm{d}W.\,^{1} \tag{A.17}$$

Recombination of electrons and holes is the cause for the reverse current and so the reverse current density J_{rad} (unit $\mathrm{A/m^2}$)for radiative recombination can be calculated to

$$\begin{aligned} J_{\mathrm{rad}} = er &= \exp\left(\frac{U}{U_T}\right) \int_{W_G}^{\infty} \frac{2\pi W^2}{c^2 h^3} \frac{1}{\exp\left(\frac{W}{kT}\right)-1}\,\mathrm{d}W \\ &= e\frac{2\pi}{c^2 h^3} \exp\left(\frac{U}{U_T}\right) \exp\left(-\frac{W_G}{kT}\right)\left[kTW_G^2 + 2(kT)^2 W_G + 2(kT)^3\right] \\ &\approx \underbrace{e\frac{2\pi}{c^2 h^3} \exp\left(-\frac{W_G}{kT}\right) kTW_G^2}_{=J_{s,\mathrm{rad}}} \exp\left(\frac{U}{U_T}\right) \end{aligned} \tag{A.18}$$

with saturation current density $J_{s,\mathrm{rad}}$. This minimum reverse current is unavoidable in p-n junctions. Calculating the reverse current $I_{s,\mathrm{rad}} = J_{s,\mathrm{rad}} F$ for an area $F = (40\,\mu\mathrm{m})^2$ using Eq. (A.18) yields 4.4 fA. Standard InGaAs pin photodiodes [94] with 40 µm in diameter show a reverse current of 30 pA. Non-radiative recombination processes which were not taken into account in the above calculations lead to a reverse current of real devices which is four orders of magnitude larger than the theoretical minimum.

---

[1] This formula for the recombination rate is an approximation which only holds when the Boltzmann approximation applies. Here, in the case of photodiodes employed as photovoltaic converters, this a very good approximation. Details and an exact derivation can be found in [1].

A.2. Impedance Matching

Impedance matching means that the maximum power is extracted from a given source. In Fig. A.1 two sources are depicted for comparison: a real current source (a) and an idealized photodiode (b). The real current source consists of an ideal current source with an internal resistor R_i connected in parallel. The source current I_q splits up into the loss current I_{Ri} through R_i and the current through the attached load I_{RL}. In the idealized photodiode instead of an internal resistor a diode is connected parallel to the ideal current source. The diode represents the p-n junction in which light is absorbed and the photo current is generated. In the equivalent circuit of the photodiode the source current I_q equals the generated photo current.

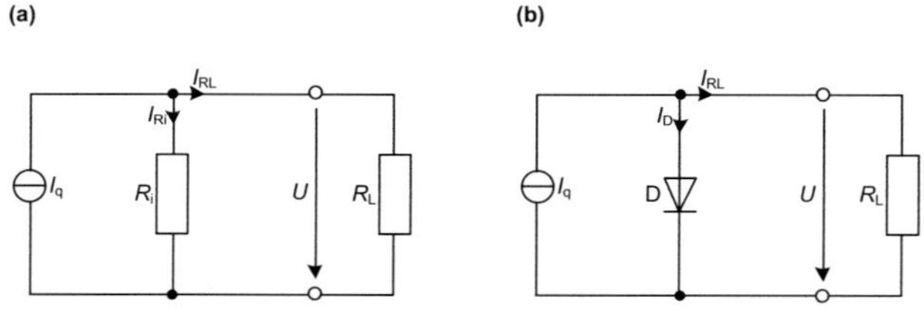

Fig. A.1 Equivalent circuits of a real current source and an idealized photodiode. (a) The real current source consists of an ideal current source with an internal resistor R_i in parallel. The source current I_q splits up into the loss current I_{Ri} through R_i and the current through the attached load I_{RL}. (b) In the idealized photodiode instead of an internal resistor a diode is connected parallel to the ideal current source.

The loss mechanisms in these sources – the internal resistor in the real current source and the diode in the photodiode – limit the power efficiency of the sources. Efficiency η is defined as ratio between power P_{RL} given to the load and total power P_{tot}

$$\eta = \frac{P_{RL}}{P_{tot}} = \frac{P_{RL}}{P_{RL} + P_{Ri}} \tag{A.19}$$

The output voltage and current depend on the attached load. Short circuiting ($R_L = 0$) gives an output voltage of 0 and the source current is flowing completely through the load. For an open circuit ($R_L = \infty$) the whole current flows internally and the voltage at the output gets maximum.

The output current in dependence of the output voltage U for the real current source is shown in Fig. A.2 (a). At a real current source the output current I_{RL} drops linearly with the voltage whereas the internal loss current I_{Ri} increases. The calculated output power dissipated in the load P_{RL} and in the internal resistor P_{Ri} calculated from (a) over output voltage U are shown in Fig. A.2 (c). The power given to the load shows a maximum where both powers are equal. This means efficiency is only 50%. The total emitted power of the current source P_{tot} is also depicted and grows linearly with the voltage.

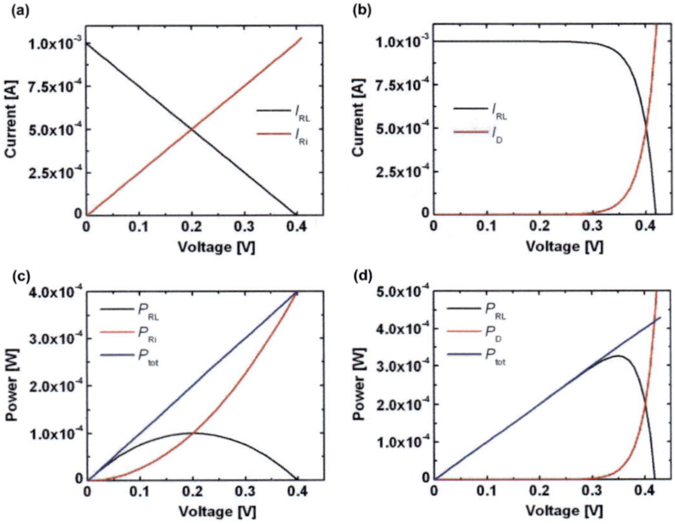

Fig. A.2 Output current and power in dependence of the output voltage for the circuits shown in Fig. A.1. (a) At a real current source the output current I_{RL} drops linearly with the voltage whereas the internal loss current I_{Ri} increases. (b) At a photodiode the exponential dependence between voltage and current of the pn-junction I_D is visible. (c) Calculated power dissipated in the load P_{RL} and in the internal resistor P_{Ri}, calculated from (a). The power given to the load shows a maximum where both powers are equal. This means efficiency is only 50%. The total emitted power of the current source P_{tot} is also depicted and grows linearly with the voltage. (d) In the photodiode also exists a maximum for the given power P_{RL} to the load. But here, efficiency is significantly higher than in comparison to (c).

In Fig. A.2 (b) the output current in dependence of the output voltage U for the idealized photodiode is shown. The exponential dependence between voltage and current through the p-n junction I_D is visible. In the photodiode exists also a maximum for the given power P_{RL} to the load which is depicted in Fig. A.2 (d). But here, efficiency is significantly higher than in comparison to (c). Reason for this can found in the current voltage characteristic of a diode. A forward biased diode needs some few 100 mV until a current can efficiently flow. This can be seen as kink in the current-voltage plot. Operated beneath the kink the output current of the photodiode stays constant whereas the voltage grows linearly. As a consequence, the photodiode is a very efficient power source.

A.3. Demarcation Device

Schematics and layouts of the battery powered demarcation devices which were discussed in Section 4.1 are presented here. Two realizations were discussed: one with a VCSEL and one with a DFB laser in the transmitter.

Schematics

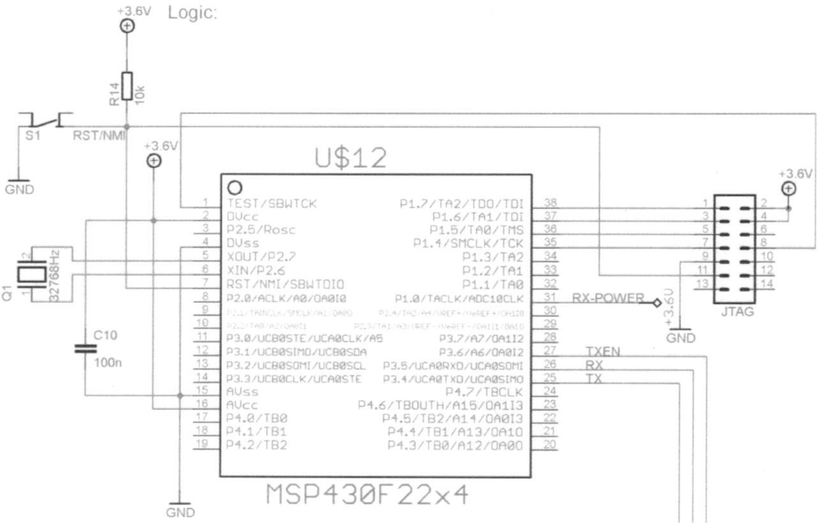

Fig. A.3 Logic circuits consisting of microcontroller (MSP430F22x4), Joint Test Action Group (JTAG) programming and debugging interface, and manual reset switch S1. Circuits used in Section 4.1.

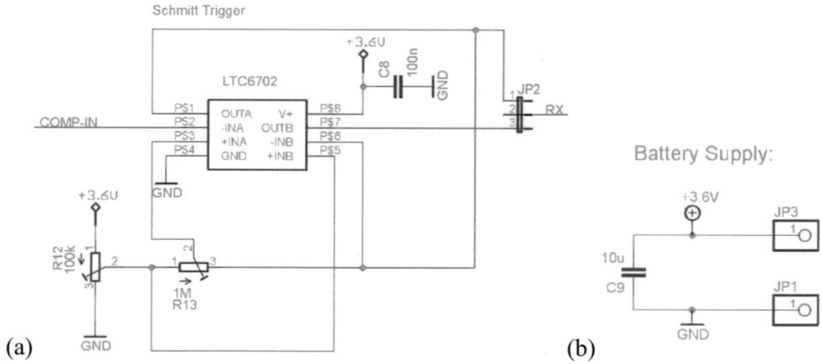

Fig. A.4 (a) Schmitt-Trigger as comparator. Two potentiometers allow individual adjustment of hysteresis and threshold. (b) Connectors for the battery. Both circuits used in Section 4.1.

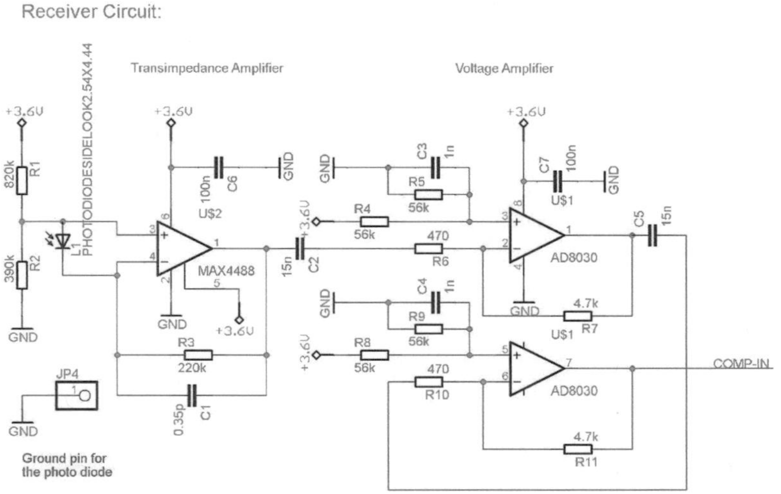

Fig. A.5 Receiver consisting of photodiode, transimpedance amplifier and dual voltage amplifier. The output signal of the voltage amplifier is routed to the comparator input (COMP-IN). Both circuits used and discussed in more detail in Section 4.1.

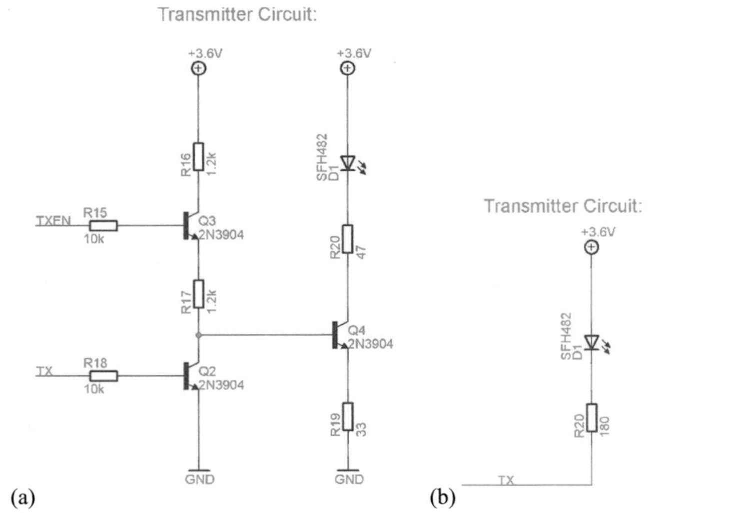

Fig. A.6 Transmitter with driver stage for (a) the DFB laser and (b) the VCSEL. Both circuits used and discussed in more detail in Section 4.1.

Layouts

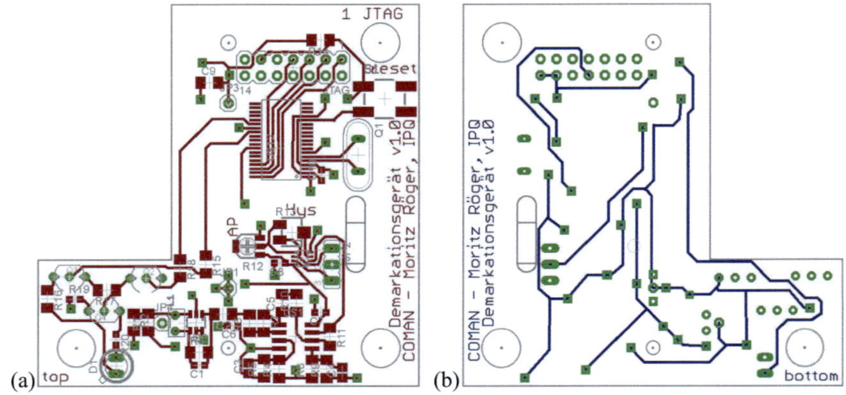

Fig. A.7 Layout of demarcation device electronics with a DFB laser as transmitter in (a) top and (b) bottom view, circuits shown in Fig. A.3 - Fig. A.6.

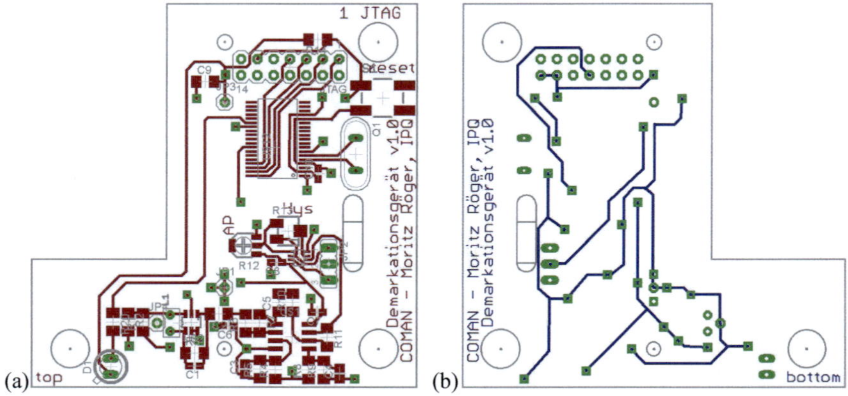

Fig. A.8 Layout of demarcation device electronics with a VCSEL as transmitter in (a) top and (b) bottom view, circuits shown in Fig. A.3 - Fig. A.6.

A.4. Optically Powered Sensor Node

Schematic and layout of the optically powered sensor node which was discussed in Section 4.2 are listed here.

Schematic

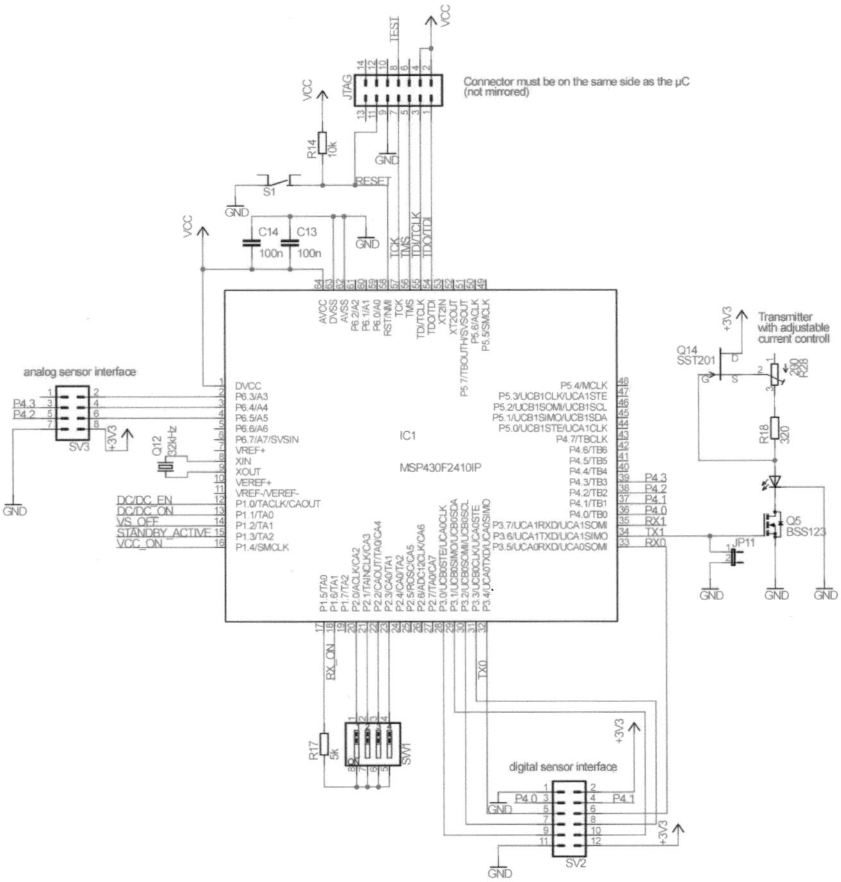

Fig. A.9 Logic circuits consisting of microcontroller (MSP430F2410IP), JTAG programming and debugging interface, driver stage for the transmitter, digital sensor interface, dual inline switch, analog sensor interface, and manual reset switch S1. Circuits used in Section 4.2.2.

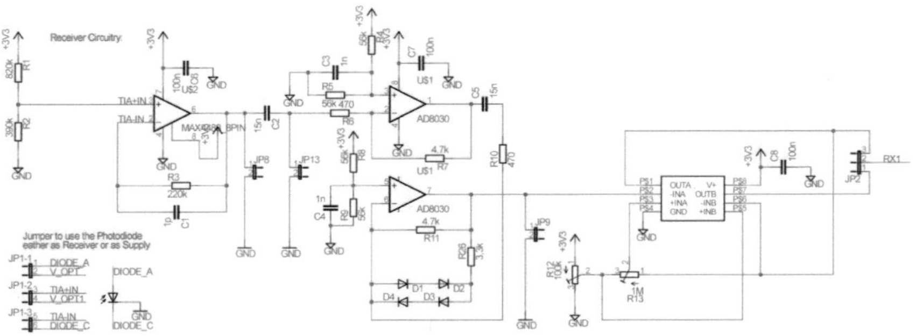

Fig. A.10 Receiver consisting of selectable photodiode, transimpedance amplifier, dual voltage amplifier with gain suppression, and comparator. Circuits used in Section 4.2.2.

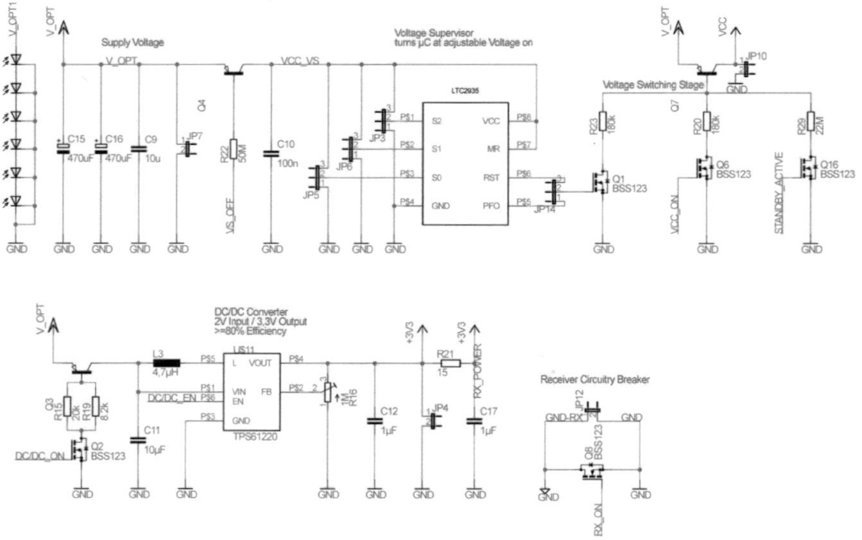

Fig. A.11 Power supply consisting of photodiode array, voltage supervisor, switches for power supplies, and DC/DC boost converter. Circuits used in Section 4.2.2.

Layout

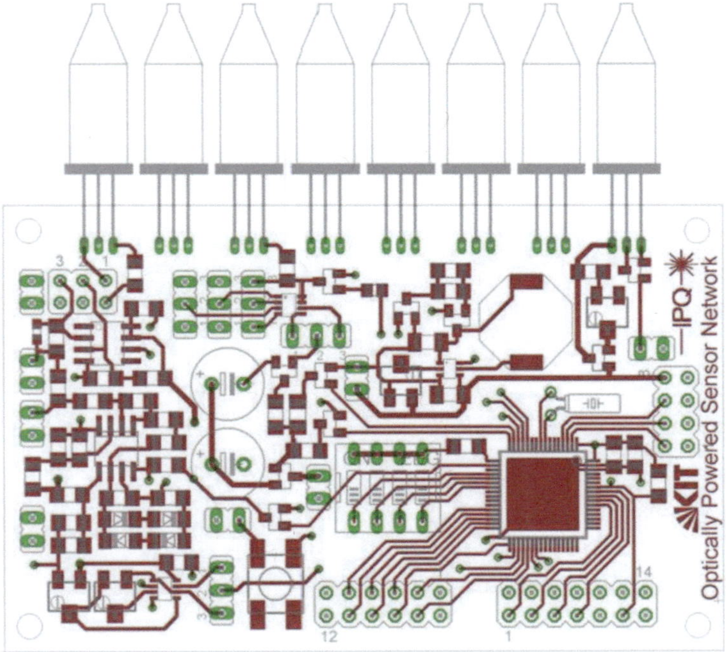

Fig. A.12 Layout of the sensor node electronics in top view, circuits shown in Fig. A.9 - Fig. A.11.

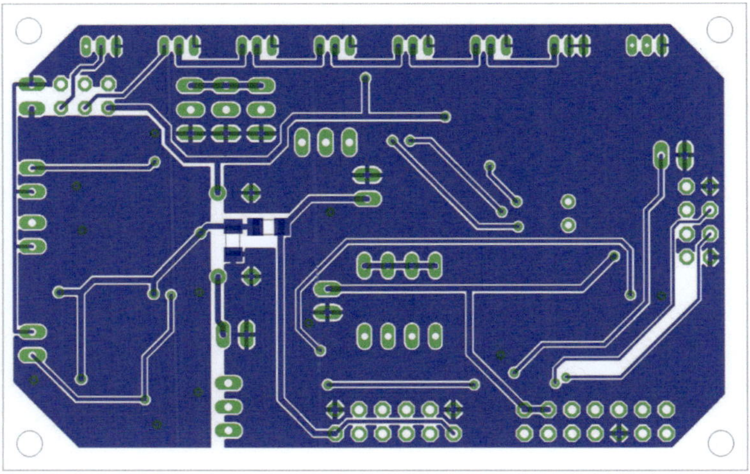

Fig. A.13 Layout of the sensor node electronics in bottom view, circuits shown in Fig. A.9 - Fig. A.11.

A.5. Constants

$$h = 6.626\,068 \cdot 10^{-34}\,\mathrm{Js} \tag{A.19}$$

$$e = 1.60217646 \cdot 10^{-19}\,\mathrm{C} \tag{A.19}$$

$$\varepsilon_0 = 8.854\,188 \cdot 10^{-12}\,\frac{\mathrm{As}}{\mathrm{Vm}} \tag{A.19}$$

$$c = \frac{1}{\sqrt{\varepsilon_0 \mu_0}} = 299\,792\,458\,\frac{\mathrm{m}}{\mathrm{s}} \tag{A.19}$$

$$k = 1.380\,6488 \cdot 10^{-23}\,\frac{\mathrm{J}}{\mathrm{K}} \tag{A.20}$$

Glossary

Acronyms

AC	Alternating current
ACK	Acknowledge
ADC	Analog-to-digital converter
Addr	Addressing interval
BS	Base station
cAWG	Cyclic arrayed waveguide grating
CMOS	Complementary metal-oxide semiconductor
CO	Central office
Com	Communication interval
CPLD	Complex programmable logic device
CRC	Cyclic redundancy check
CSMA	Carrier sense multiple access
CW	Continuous-wave
DC	Direct current
DCO	Digitally-controlled oscillator
DD	Demarcation device
DFB	Distributed feedback
FER	Frame error ratio
FET	Field-effect transistor
FPGA	Field programmable gate array
FTT*x*	Fiber-to-the-*x* systems with x = C, B, H for cabinet, building, home
GI	Gradient-index
GSM	Groupe spécial mobile
GPON	Gigabit passive optical network
high-Q	High-quality
HP	High power
I^2C	Inter-integrated circuit
IO	Input/output
ISO	International organization for standardization
JPEG	Joint photographic experts group
JTAG	Joint Test Action Group (IEEE standard 1149.1)
LD	Laser diode
LE-MAC	Low-energy MAC
low-Q	Low-quality
LPM	Low power modes
Lstn	Listen interval

LTE	Long term evolution
MAC	Media access control
MEMS	Micro-electromechanical system
MMF	Multimode fiber
NPN	Bipolar junction transistor with doping profile n-p-n
NRZ	Non-return to zero
O/E	Opto-electrical
ODN	Optical distribution network
OLT	Optical line termination
ONU	Optical network unit
ONT	Optical network termination
OP	Operational amplifier
OTDR	Optical time domain reflectometry
PCB	Printed circuit board
PD	Photodiode
PNP	Bipolar junction transistor with doping profile p-n-p
PON	Passive optical network
PPC	Photovoltaic power converter
PRBS	Pseudo random bit sequence
PV	Photovoltaic
PVC	PV converter
RAM	Random access memory
RF	Radio frequency
Rp	Photonic-power receiver
Rq	Request
RV	Rendezvous sequence
Rx	Receiver
S-MAC	Sensor MAC
SCP-MAC	Scheduled channel polling MAC
SMF	Singlemode fiber
SNR	Signal to noise ratio
SoC	System-on-chip
SPI	Serial peripheral interface
TDM	Time division multiplexing
TIA	Transimpedance amplifier
Tx	Transmitter
TxEn	Transmitter enable
UART	Universal asynchronous receiver / transmitter
UMTS	Universal mobile telecommunications system
USB	Universal serial bus
VCSEL	Vertical cavitiy surface emitting laser
VGA	Video graphics array

VLO	Very-low-power low-frequency oscillator
VoIP	Voice-over-internet protocol
VS	Voltage supervisor
VV	Voltage amplifer
WDM	Wavelength division multiplexing
WkUp	Waking up
WLAN	Wireless local area network
WSN	Wireless sensor network
WWW	World wide web
YCbCr	Luma component, blue-difference and red-difference chroma component.
μC	Microcontroller

References

[1] P. Würfel, *Physics of Solar Cells*. Weinheim: Wiley-VCH, 2009.

[2] R. C. Miller, B. Schwartz, L. A. Koszi, and W. R. Wagner, "A high-efficiency GaAlAs double-heterostructure photovoltaic detector," *Appl. Phys. Lett.*, vol. 33, pp. 721–723, 1978.

[3] A. W. Bett, F. Dimroth, R. Löckenhoff, E. Oliva, J. Schubert, "III-V solar cells under monochromatic illumination," in *Proc. 33rd IEEE Photovoltaic Spec. Conf.*, San Diege, CA, USA, May 12–16, pp. 1–5, 2008.

[4] B. C. DeLoach, R. C. Miller, and S. Kaufman, "Sound alerter powered over an optical fiber," *Bell. Syst. Tech. J.*, vol. 57, pp. 3309–3316, 1978.

[5] R. C. Miller and R. B. Lawry, "Optically powered speech communication over a fiber lightguide," *Bell. Syst. Tech. J.*, vol. 58, pp. 1735–1741, 1979.

[6] R. C. Miller, B. C. DeLoach, T. S. Stakelon, and R. B. Lawry, "Wideband, bidirectional lightguide communication with an optically powered audio channel," *Bell Syst. Tech. J.*, vol. 61, pp, 1359–1365, 1982.

[7] H. Kirkham and A. R. Johnston, "Optically powered data link for power system applications," *IEEE Trans. Power Delivery*, vol. 4, pp. 1997–2004, 1989.

[8] T. C. Banwell, R. C. Estes, L. A. Reith, P. W. Shumate, Jr., and E. M. Vogel, "Powering the fiber loop optically – A cost analysis," *J. Lightwave Technol.*, vol. 11, pp. 481–494, 1993.

[9] S. J. Pember, C. M. France, and B. E. Jones, "A multiplexed network of optically powered, addressed and interrogated hybrid resonant sensors," *Sens. Actuators A*, vol. 47, pp. 474–477, 1995.

[10] M. Q. Feng, "Optically powered electrical accelerometer and its field testing," *J. Eng. Mech.*, vol, 124, pp. 513–519, 1998.

[11] R. Pena, C. Algora, I. R.Matías, and M. López-Amo, "Fiber-based 205-mW (27% efficiency) power-delivery system for an all-fiber network with optoelectronic sensor units," *Appl. Opt.*, vol, 38, pp. 2463–2466, 1999.

[12] H. Miyakawa, Y. Tanaka, and T. Kurokawa, "Design approaches to power-over-optical local-area-network systems," *Appl. Opt.*, vol. 43, pp. 1379–1389, 2004.

[13] G. Böttger, M. Dreschmann, C. Klamouris, M. Hübner, M. Röger, A. W. Bett, T. Kueng, J. Becker, W. Freude, and J. Leuthold, "An Optically Powered Video Camera Link," *IEEE Photon. Technol. Lett.*, vol. 20, pp. 39–41, January 1, 2008.

[14] K. Liu, "Power budget considerations for optically activated conventional sensors and actuators," *IEEE Trans. Instrum. Meas.*, vol. 40, pp. 25–27, 1991.

[15] S. van Riesen, U. Schubert, and A. W. Bett, "GaAs photovoltaic cells for laser power beaming at high power densities," in *Proc. 17th Eur. PV Solar Energy Conf.*, Munich, Germany, 2001, pp. 18–21, paper VA1/26.

[16] H. Miyakawa, Y. Tanaka, and T. Kurokawa, "Photovoltaic cell characteristics for high-intensity laser light," *Sol. Energy Mater. Sol. Cells*, vol. 86, pp. 253–267, 2005.

[17] J. G. Werthen, "Powering next generation networks by laserlight over fiber," in *Proc. Optical Fiber Communication Conference / National Fiber Optic Engineers Conference, 2008. OFC/NFOEC 2008*, San Diego, CA, USA, Feb. 24–28, 2008, paper OWO3.

[18] S. M. Sze, *Physics of Semiconductor Devices*. Hoboken: John-Wiley & Sons, 2007.

[19] N. W. Ashcroft, N. D. Mermin, *Solid State Physics*. Belmont: Brooks/Coole Cengage Learning, 2009.

[20] C. Kittel, *Introduction to Solid State Physics*. New York: Wiley, 1986.

[21] W. Shockley, H .J .Queisser, "Detailed balance limit of efficiency of p-n junction solar cells", *J. Appl. Phys.*, vol. 32, is. 3, pp. 510-519, 1961.

[22] V. M. Andreev, V. A. Grilikhes, V. D. Rumyantsev, *Photovoltaic Conversion of Concentrated Sunlight*. Chichester: John Wiley & Sons, 1997.

[23] L. C. Olsen, D. A. Huber, G. Dunham, F. W. Addis, "High efficiency monochromatic GaAs solar cells," in *Conference Record of the 22nd IEEE Photovoltaic Specialists Conference,* 7-11 Oct, 1991, pp.419-424.

[24] "Optics and photonics - Spectral bands, ISO Standard 20473:2007," International Organization for Standardization, Geneva, Switzerland. [Online]. Available: http://-www.iso.org/iso/iso_catalogue/catalogue_tc/catalogue_detail.htm?csnumber=39482

[25] P. G. Borden, "A monolithic series-connected $Al_{0.93}Ga_{0.07}As$/GaAs solar cell array;" *Appl. Phys. Lett.*, vol. 35, is. 7, pp. 553-554, 1979.

[26] Alberto Leon-Garcia, Indra Widjaja, *Communication Networks*. New York: McGraw-Hill, 2006.

[27] Guegan M. (Michel.GUEGAN@saftbatteries.com) "RE: Durability of batteries." E-mail to the author. 18 March 2009.

[28] C. Cano, B. Bellalta, A. Sfairopoulou, M. Oliver, "Low energy operation in WSNs: A survey of preamble sampling MAC," *Comput. Netw.*, vol. 55, is. 15, pp. 3351–3363, July 2011.

[29] W. Ye, J. Heidemann, D. Estrin, "An energy-efficient MAC protocol for wireless sensor networks," in *Proceedings of the 21st Annual Joint Conference of the IEEE Computer and Communications Societies*, vol. 3, pp. 1567–1576, 2002.

[30] J. Hill, D. Culler, "Mica: a wireless platform for deeply embedded networks," *IEEE Micro*, pp. 12–24, 2002.

[31] A. El-Hoiydi, "Aloha with preamble sampling for sporadic traffic in Ad hoc wireless sensor networks," in *Proceedings of the IEEE International Conference on Communications (ICC 02)*, vol. 5, pp. 3418–3423, 2002.

[32] S. Lim, S. Kim, J. Cho, S. An, "Medium access control with an energy-efficient algorithm for wireless sensor networks," in *Proceedings of the Personal Wireless Communications*,". pp. 334–343, 2006.

[33] X. Shi, G. Stromberg, Y. Gsottberger, T. Sturm, "Wake-Up-Frame Scheme for Ultra Low Power Wireless Transceivers," in *Proceedings of GLOBECOM*, Nov. 2004.

[34] W. R. Heinzelman, A. Chandrakasan, H. Balakrishnan, "Energy-efficient communication protocol for wireless microsensor networks," in *Proc. 33rd Hawaii Int. Conf. System Sciences*, Hawaii, Jan. 4–7, vol. 2, pp. 1–10, 2000.

[35] W. Ye, F. Silva, and J. Heidemann, "Ultra-low duty cycle MAC with scheduled channel polling," in *Proc. 4th ACM SenSys Conf.*, Boulder (CO), Nov. 1–3, pp. 321–334, 2006.

[36] "MSP430™ Ultra-Low Power 16-Bit Microcontrollers," Texas Instruments Inc., Dallas, TX, USA. [Online]. Available: http://www.ti.com/lsds/ti/microcontroller/16-bit_msp430/overview.page

[37] "Microcontroller UART tutorial," Society of Robots. [Online]. Available: http://-www.societyofrobots.com/microcontroller_uart.shtml

[38] G. Böttger, M. Dreschmann, M. Röger, M. Hübner, C. Klamouris, A. W. Bett, T. Kueng, J. Becker, W. Freude, J. Leuthold, "Optically powered video camera link," in *Proc. 33th European Conf. Opt. Commun. (ECOC'07)*, Berlin, Germany, September 16–20, 2007, Thu9.6.3.

[39] "WG802.11 - Wireless LAN Working Group," IEEE Standards Association, Piscataway, NJ, USA. [Online]. Available: http://standards.ieee.org/develop/wg/-WG802.11.html

[40] "Technology Overview," Bluetooth Special Interest Group, Kirkland, WA, USA. [Online]. Available: http://developer.bluetooth.org/KnowledgeCenter/-TechnologyOverview/Pages/Technology-Overview.aspx

[41] "ZigBee Standards Overview," ZigBee Alliance, San Ramon, CA, USA. [Online]. Available: http://zigbee.org/Standards/Overview.aspx

[42] "GSA - The Global mobile Suppliers Association," Global mobile Suppliers Association, Sawbridgeworth, UK. [Online]. Available: http://www.gsacom.com/

[43] "ITU global standard for international mobile telecommunications ′IMT-Advanced′," International Telecommunication Union (ITU), Geneva, Switzerland. [Online]. Available: http://www.itu.int/ITU-R/index.asp?category=information&rlink=imt-advanced&lang=en

[44] "3GPP Specifications," 3GPP Mobile Competence Centre, Sophia-Antipolis Cedex, F. [Online]. Available: http://www.3gpp.org/specifications

[45] R. C. Miller and R. B. Lawry, "Optically powered speech communication over a fiber lightguide," *Bell Syst. Tech. J.*, vol. 58, no. 7, pp. 1735–1741, Sep. 1979.

[46] F. Caspers and E.-G. Neumann, "Optical power supply for measuring or communication devices at high-voltage levels," *IEEE Trans. Instrum. Meas.*, vol. IM-29, no. 1, pp. 73–74, Mar. 1980.

[47] J. G. Werthen, A. G. Andersson, H. O. Björklund, and S. T. Weiss, "Current measurements using optical power," *IEEE Transmission Distribution Conf.*, pp. 213–218, 1996.

[48] C. R. Giles, A. Dentai, E. Burrows, C. A. Burrus, L. Kohutich, and J. Centanni, "Microwatt-power InGaAs photogenerator for lightwave networks," *IEEE Photon. Technol. Lett.*, vol. 9, no. 5, pp. 666–668, May 1997.

[49] H. Ramanaitra, P. Chanclou, Z. Belfqih, M. Moignard, H. Le Bras, and D. Schumacher, "Scalable and multi-service passive optical access infrastructure using variable optical splitters," in *Optical Fiber Communication Conference, 2006 and the 2006 National Fiber Optic Engineers Conference. OFC 2006*, Anaheim, CA, USA, Mar. 5–10, 2006, paper OFE2.

[50] Y. Liu, J. J. Brown, D. C. W. Lo, and S. R. Forrest, "Optically powered optical interconnection system," *IEEE Photon. Technol. Lett.*, vol. 1, no. 1, pp. 21–23, Jan. 1989.

[51] H. Miyakawa, Y. Tanaka, and T. Kurokawa, "Design approaches to power-over-optical local-area-network systems," *Appl. Opt.*, vol. 43, no. 6, pp. 1379–1389, 2004.

[52] H. Miyakawa, Y. Tanaka, and T. Kurokawa, "Optical power and signal transmission with WDM single-mode fiber optic systems," in *Proc. Renewable Energy Conf.*, pp. 317–320, 2006.

[53] R. Pena, C. Algora, I. R. Matias, and M. Lopez-Amo, "Fiber-based 205-mW (27% efficiency) power-delivery system for an all-fiber network with optoelectronic sensor units," *Appl. Opt.*, vol. 38, no. 12, pp. 2463–2466, Apr. 1999.

[54] A G. Dentai, C. R. Giles, E. Burrows, C. A. Burrus, L. Stulz, J. Centanni, J. Hoffman, and B. Moyer, "A long-wavelength 10-V optical-to-electrical InGaAs photo–generator," *IEEE Photon. Technol. Lett.*, vol. 11, no. 1, pp. 114–116, Jan. 1999.

[55] ITU-T Recommendation G.984.2: "Gigabit-capable passive optical networks (GPON): physical media dependent (PMD) layer specification," March 2003

[56] H.-C. Lu and W.-S. Wang, "Cyclic arrayed waveguide grating device with flap-top passband and uniform spectral response," *IEEE Photon. Technol. Lett.*, vol. 20, no. 1, pp. 3–5, Jan. 2008.

[57] Y. Inoue, A. Himeno, K. Moriwaki, and M. Kawachi, "Silica-based arrayed-waveguide grating circuit as optical splitter/router," *Electron. Lett.*, vol. 31, no. 9, pp. 726–727, Apr. 1995.

[58] J. Mu, C. Xu, and W.-P. Huang, "An optical power combiner/wavelength demultiplexing module for hybrid WDM FTTX," *Opt. Express*, vol. 17, no. 6, pp. 4791–4797, March 2009.

[59] K. Tanaka, M. Tateda, and Y. Inoue, "Measuring the individual attenuation distribution of passive branched optical networks," *IEEE Photon. Technol. Lett.*, vol. 8, no. 7, pp. 915–917, July 1996.

[60] "TPS61201 Low Input Voltage Synchronous Boost Converter With 1.3-A Switches, Datasheet," Texas Instruments Incorporated, Dallas, TX , 2008

[61] "LTC3105 - 400mA Step-Up DC/DC Converter with Maximum Power Point Control and 250mV Start-Up," Datasheet, Linear Technology Corporation, Milpitas, CA, 2010

[62] "LTC2935 - Ultra-Low Power Supervisor with Power-Fail Output, Selectable Thresholds," Datasheet, Linear Technology Corporation, Milpitas, CA, 2008. [Online]. Available: http://cds.linear.com/docs/Datasheet/2935fa.pdf

[63] R. C. Miller, B. Schwartz, L. A. Koszi, and W. R. Wagner, "A high-efficiency GaAlAs double-heterostructure photovoltaic detector," *Appl. Phys. Lett.*, vol. 33, no. 8, pp. 721–723, Oct. 1978.

[64] S. van Riesen, U. Schubert, and A. W. Bett, "GaAs photovoltaic cells for laser power beaming at high power densities," in *Proc. 17th Eur. PV Solar Energy Conf.,* Munich, Germany, 2001, pp. 18–21, paper VA1/26.

[65] C. Klamouris, G. Boettger, M. Huebner, M. Dreschmann, K. Paulsson, A. W. Bett, T. Kueng, J. Becker, W. Freude, and J. Leuthold, "Optically powered platform with Mb/s transmission over a single fiber," in *Proc. 32th European Conf. Optical Communications (ECOC'06),* Cannes, France, September 2006, vol. 3, pp. 461–462, 2006, paper We3.P.170.

[66] "Aeroflex Gaisler AB," Aeroflex Gaisler AB, Goteborg, Sweden. [Online]. Available: http://www.gaisler.com

[67] "TPS61030: Adjustable, 4-A Switch, 96% Efficient Boost Converter w/ 20µA Iq in TSSOP-16," Datasheet, Texas Instruments Inc., Dallas, TX, USA. [Online]. Available: http://www.ti.com/lit/ds/slus534e/slus534e.pdf

[68] "TPS62000: Adjustable, 600-mA, 95% Efficient Step-Down Converter," Datasheet, Texas Instruments Inc., Dallas, TX, USA. [Online]. Available: http://www.ti.com/lit/-ds/symlink/tps62000.pdf

[69] "The ultra-low-power programmable solution," Microsemi SoC Products Group, San Jose, CA, USA. [Online]. Available: http://www.actel.com/products/igloo/default.aspx

[70] "OmniVision," OmniVision, Santa Clara, CA, USA. [Online]. Available: http://-www.ovt.com/products/

[71] "Product Specification Virtex®-5 family," Xilinx, Inc., San Jose, CA, USA. [Online]. Available: http://www.xilinx.com/support/documentation/data_sheets/ds100.pdf

[72] M. Röger, G. Böttger, M. Dreschmann, C. Klamouris, M. Huebner, A. W. Bett, J. Becker, W. Freude, and J. Leuthold, "Optically powered fiber networks, " *Opt. Express*, vol. 16, is. 26, pp. 21821–21834, Dec. 2008, Focus Issue on Optics for Energy [invited].

[73] Th. Pfeiffer, J. Hehmann, H. Schmuck, W. Freude, J. Vandewege, and H. Yanagisawa, "Monitoring and protecting the optical layer in FTTH networks," in *Proc. FTTH Conf. & Expo,* Las Vegas, NV, USA, Oct. 3–6, 2005.

[74] M. Freiberger and S. Huq, "Network element security in optical communications," in *Proc. Optical Fiber Communication Conference / National Fiber Optic Engineers Conference, 2008. OFC/NFOEC 2008*, San Diego, CA, USA, Feb. 24–28, 2008, paper NWA1.

[75] C. C. K. Chan, L.-K. Chen, C. Lin, "Novel network architectures for survivable WDM passive optical," in *Proc. 34th European Conf. Optical Communications (ECOC'08),* Brussels, Belgium, Sep. 21–25, 2008, paper Th.1.F.6.

[76] Th. Pfeiffer, H. Schmuck, M Straub, J. Hehmann, "Cost efficient non-service interrupting monitoring of optical fiber links in FTTH / FTTB networks," in *Proc. 34th European Conf. Optical Communications (ECOC'08),* Brussels, Belgium, Sep. 21–25, 2008, paper Tu.1.F.2.

[77] M. Röger, M. Hoh, J. Hehmann, T. Pfeiffer, M. Hübner, J. Becker, J. Leuthold, and W Freude, "Low-energy demarcation device for monitoring FTTx networks," in *Proc. 10. VDE-ITG-Fachtagung Photonische Netze*, Deutsche Telekom Hochschule für Telekommunikation, Leipzig, May 2009.

[78] J. Hehmann and Th. Pfeiffer, "New monitoring concepts for optical access networks," *Bell Labs Technical Journal*, vol. 13, is. 1, pp. 183–198, May 2008.

[79] *Cadence OrCAD*, Cadence Design Systems, Berkshire, United Kingdom. [Online]. Available: http://www.cadence.com/products/orcad/pages/default.aspx

[80] H. T. Friis, "Noise Figures of Radio Receivers," *Proceedings of the IRE*, vol. 32, no. 7, pp. 419- 422, July 1944

[81] W. Freude, *Optical Communications Systems – Part 2: Receivers*, Institute of Photonics und Quantum Electronics (IPQ), Karlsruhe Institute of Technology, Nov. 2011 [Online]. Available: http://www.ipq.kit.edu/downloads/ocs-rxpdf.zip

[82] "Datasheet 1550nm VCSEL," RayCan, Daejon, Korea. [Online]. Available: http://-www.raycan.com/image/pdf/RC32xxx1-T.pdf

[83] "Wireless Aplications," Part Number DFB-1490-C5-2-2.5-FA-A-A-I, Applied Optoelectronics, Inc., Sugar Land, TX, USA. [Online]. Available: http://www.ao-inc.com/products/wireless.php

[84] C. Lange, D. Kosiankowski, R. Hülsermann, A. Gladisch, R. Weidmann, "Energy Footprint of Telecommunication Networks," in *Proc. 36th European Conf. Optical Communications (ECOC'10),* Torino, Italy, Sep. 19–23, 2010, paper Mo.1D.2.

[85] S.-W. Wong , S.-H. Yen, P. Afshar, S. Yamashita, and L. G. Kazovsky, "Demonstration of Energy Conserving TDM-PON with Sleep Mode ONU using Fast Clock Recovery Circuit," in *Proc. Optical Fiber Communication Conference / National Fiber Optic Engineers Conference, 2010. OFC/NFOEC 2010*, San Diego, CA, USA, Mar. 21–25, 2010, paper OThW7.

[86] "2SA1338/2SC3392 – PNP/NPN Epitaxial Planar Silicon Transistors," Datasheet, Sanyo Electric Co., Tokyo, Japan. [Online]. Available: http://-www.datasheetcatalog.org/datasheet/sanyo/ds_pdf_e/2SC3392.pdf

[87] "FDV301N N-Channel, Digital FET," Datasheet, Fairchild Semiconductor Corporation, San Jose, CA, USA. [Online]. Available: http://www.fairchildsemi.com/-ds/FD/FDV301N.pdf

[88] "Elna," Elna America, Inc., Cypress, CA, USA. [Online]. Available: http://www.elna-america.com/

[89] "DS1722 Digital Thermometer with SPI/3-Wire Interface," Datasheet, Maxim Integrated Products, Inc., Sunnyvale, CA, USA. [Online]. Available: http://-datasheets.maxim-ic.com/en/ds/DS1722.pdf

[90] "MEMSIC," MEMSIC Inc., Andover, MA, USA. [Online]. Available: http://-www.memsic.com/

[91] "APDS-9003 Miniature Surface Mount Ambient Light Photo Sensor," Avago Technologies, San Jose, CA, USA. [Online]. Available: http://www.avagotech.com/-pages/en/optical_sensors/ambient_light_photo_sensors/apds-9003/

[92] K. C. Budka, J. G. Deshpande, T. L. Doumi, M. Madden, and T. Mew, "Communication network architecture and design principles for smart grids," *Bell Labs Tech. J.*, vol. 15, is. 2, pp. 205–227, Sep. 2010.

[93] M. Roeger, B. Hiba, M. Hoh, J. Hehmann, T. Pfeiffer, M. Huebner, J. Becker, J. Leuthold, and W. Freude, "Optically powered low-energy demarcation device for monitoring FTTx networks," in *Proc. Optical Fiber Communication Conference / National Fiber Optic Engineers Conference, 2010. OFC/NFOEC 2010*, San Diego, CA, USA, Mar. 21–25, 2010, paper NWC4.

[94] "Indium Gallium Arsenide Detectors," Teledyne Judson Technologies, Montgomeryville, PA, USA. [Online]. Available: http://–www.judsontechnologies.com/files/pdf/InGaAs_shortform_DEC2004_rev2.pdf

Acknowledgements (German)

Den zahlreichen Personen, die an der erfolgreichen Fertigstellung dieser Arbeit direkt oder indirekt beteiligt waren, möchte ich an dieser Stelle meinen großen und aufrichtigen Dank für ihre vielfältige Unterstützung aussprechen.

An erster Stelle gebührt mein Dank Herrn Prof. Dr.-Ing. Dr. h. c. Wolfgang Freude. Sein mir entgegengebrachtes Vertrauen und Interesse, seine breite wie tiefe Fachkompetenz und sein unermüdlicher Einsatz beim sorgfältigen Lesen und Überarbeiten von diversen Manuskripten waren immer Stütze meiner Arbeit.

Herrn Prof. Dr. sc. nat. Jürg Leuthold danke ich für seine überblickende Sicht auf die Dinge und für die Übernahme des Korreferats.

Prof. Dr.-Ing. Jürgen Becker danke ich für die Übernahme des zweiten Korreferats und für sein Engagement bei der Entwicklung unseres gemeinsamen Projekts „Optisch versorgte Videokamera".

Für die gute und erfolgreiche Zusammenarbeit möchte ich meinen Industrie-Partnern aus dem COMAN-Projekt danken.

Bei den Bell-Labs, Alcatel-Lucent in Stuttgart seien hier Herr Dr. Thomas Pfeiffer als Leiter der Gruppe „Optische Zugangsnetzwerke", sowie deren fleißige Mitglieder Jörg Hehmann, Michael Straub und Harald Schmuck erwähnt. Viel Wissen über die letzte Meile konnte ich aus ihrem Labor entführen.

Für die interessanten Einblicke in die Welt von Fasern, Steckverbindern und anderen Faserverbindungstechniken bedanke ich mich bei Herrn Matthias Hedrich, Kilian Halbe und Christoph Raab von Diamond.

Meinen Projektpartner Dr. Thomas Paatzsch und Ingo Smaglinski von Cube Optics danke ich für die gute Kooperation im Bereich ihrer miniaturisierten Freistrahloptiken.

Prof. Dr. Hübner als meinem ehemaligen primären Ansprechpartner vom Institut für Technik der Informationsverarbeitung (ITIV) danke ich für seine Funktion als Wegbereiter für die FPGAs in unsere optisch versorgten Sensoren und für den Zugriff auf die ITIV-Werkstatt.

Dr. Andreas Bett vom Fraunhofer-Institut für Solare Energiesysteme in Freiburg danke ich für die Überlassung von zahlreichen photovoltaischen Konvertern, die unsere Sensorknoten verlässlich mit Strom versorgt haben.

Mein beiden Bürokollegen Swen König und Jingshi Li danke für die entspannte Stimmung im Büro und die gute Arbeitsatmosphäre, die bei uns Tag wie Nacht herrschte.

Meinen Studenten Moritz Baier, Florian Boes, Michael Dreschmann, Raoul Fischer, Benian Hiba, Matthias Hoh, Stephan Huber, Andreas Kleff, Martin Reichel, Eduard Tietz, Benjamin Volzer, Matthias Wippler und Ruiwen Yang bin ich zu großem Dank verpflichtet, da sie als Diplom-, Studien-, Bachelorarbeiter und als Hiwi wertvolle Beiträge geleistet haben.

Unserer feinmechanischen Werkstatt unter Leitung von Hans Bürger mit Manfred Hirsch, Werner Höhne und ihren Azubis danke ich für die stets zügige und unkomplizierte Zusammenarbeit – auch wenn die erteilten Aufträge nicht immer dienstlichen Charakter hatten. Des Weiteren gebührt Oswald Speck mein Dank für den Bau von Gehäusen für photovolataische Konverter und das geduldige Umspulen diverser Fasern. Werner Podszus und Martin Winkeler danke ich für die guten Diskussion um und Arbeiten an elektronischen Schaltungen. Bernadette Lehmann danke ich für die Erledigung zahlreicher administrativer Aufgaben, aber auch für Kaffee, Leckereien und den ein oder anderen Scherz für den sie immer zu haben war. Auch ihren Kolleginnen Ilse Kober, Angelika Olbrich, Eva-Maria Schubart und Andrea Riemensperger möchte ich an dieser Stelle meinen Dank aussprechen. Johann Hartwig Hauschild und Sebastian Struck danke ich für alle Arbeiten rund um die notwendige IT. Für das schnelle und unkomplizierte Ätzen diverser Platinen danke ich Ronald Vester vom IHE.

Für die gute Zusammenarbeit danke ich all meinen IPQ-Kollegen Luca Alloatti, David Hillerkuss, Christos Klamouris, Dr. Sebastian Köber, Swen König, Dietmar Korn, Matthias Lauermann, Jingshi Li, Nicole Lindenmann, Alexandra Ludwig, Argishti Melikyan, Dr. Arvind Mishra, Sascha Mühlbrandt, Dr. Sean O'Duill, Robert Palmer, Jörg Pfeifle, Philipp Schindler, René Schmogrow, Simon Schneider, Claudius Weimann und Kai Worms sowie meinen ehemaligen Kollegen Dr. René Bonk, Dr. Jan Brosi, Dr. Gunnar Böttger, Dr. Thomas Vallaitis, Prof. Dr. Christian Koos, Dr. Philipp Vorreau und Andrej Marculescu.

Für die gute Zeit am aber auch außerhalb vom IPQ möchte ich Swen und René danken.

List of Publications

Journal Papers

[J1] **M. Roeger**, B. Hiba, J. Hehmann, M. Straub, H. Schmuck, M. Hedrich, T. Pfeiffer, C. Koos, J. Leuthold, W. Freude, "In-service Monitoring of PON Access Networks with Powerline Independent Devices," to be submitted.

[J2] D. Hillerkuss, R. Schmogrow, T. Schellinger, M. Jordan, M. Winter, G. Huber, T. Vallaitis, R. Bonk, P. Kleinow, F. Frey, **M. Roeger**, S. Koenig, A. Ludwig, A. Marculescu, J. Li, M. Hoh, M. Dreschmann, J. Meyer, S. Ben Ezra, N. Narkiss, B. Nebendahl, F. Parmigiani, P. Petropoulos, B. Resan, A. Oehler, K. Weingarten, T. Ellermeyer, J. Lutz, M. Moeller, M. Huebner, J. Becker, C. Koos, W. Freude, and J. Leuthold, "26 Tbit s−1 line-rate super-channel transmission utilizing all-optical fast Fourier transform processing," *Nature Photonics*, vol. 5, pp. 364–371, May 2011, doi: 10.1038/nphoton.2011.74.

[J3] M. Dreschmann, **M. Röger**, M. Hübner, J. Becker, W. Freude, J. Leuthold, and V. Schäfer, "Lichtblick – Intelligente Kamera-Sensoren via Glasfaser mit Strom versorgen," *elektronikJOURNAL* 06/2009, pp. 70–72.

[J4] **M. Röger**, G. Böttger, M. Dreschmann, C. Klamouris, M. Huebner, A.W. Bett, J. Becker, W. Freude, and J. Leuthold, "Optically powered fiber networks,“ *Opt. Express*, vol. 16, is. 26, pp. 21821–21834, Dec. 2008, Focus Issue on Optics for Energy [invited], doi: 10.1364/OE.16.021821.

[J5] G. Böttger, M. Dreschmann, C. Klamouris, M. Hübner, **M. Röger**, A.W. Bett, T. Kueng, J. Becker, W. Freude, and J. Leuthold, "An Optically Powered Video Camera Link,“ *IEEE Photon. Technol. Lett.*, vol. 20, pp. 39–41, January 1, 2008, doi: 10.1109/LPT.2007.912695.

[J6] **M. Röger**, M. Uhlarz, and H.v. Löhneysen, "Magnetotransport across the field-induced quantum phase transition in $CeCu_{5.8}Au_{0.2}$,“ *J. Magn. Magn. Mater.*, vol. 310, is. 2 part 1, pp. 847–848, March 2007, doi: 10.1016/j.jmmm.2006.10.873.

Conference Contributions

[C1] **M. Roeger,** F. Boes, A. Kleff, B. Hiba, M. Baier, M. Hoh, S. Koenig, C. Koos, J. Leuthold, and W. Freude, "Energy-efficient MAC Protocol Enabling an Optically Powered Sensor Network," in *Proc. Optical Fiber Communication Conference 2011,* Los Angeles, CA, USA, March 2011, paper OML4.

[C2] M. Dreschmann, M. Hübner, **M. Röger**, O. Sander, C. Klamouris, J. Becker, W. Freude, and J. Leuthold, "Reconfigurable hardware for power-over-fiber applications," in *Proc. 20th International Conference on Field Programmable Logic and Applications (FPL)*, Milano, Italy, August 2010.

[C3] W. Freude, **M. Roeger**, J. Hehmann, T. Pfeiffer, M. Huebner, J. Becker, C. Koos, and J. Leuthold, "Energy-autarkic monitor for FTTx networks," in *Proc. OSA Optics & Photonics Congress*, Karlsruhe, Germany, June 2010, paper ATuC2 [invited].

[C4] **M. Roeger**, B. Hiba, M. Hoh, J. Hehmann, T. Pfeiffer, M. Huebner, J. Becker, J. Leuthold, and W. Freude, "Optically powered low-energy demarcation device for monitoring FTTx networks," in *Proc. Optical Fiber Communication Conference / National Fiber Optic Engineers Conference (OFC/NFOEC'10)*, San Diego (CA), USA, March 2010, paper NWC4.

[C5] W. Freude, **M. Roeger**, B. Hiba, M. Hoh, M. Dreschmann, J. Hehmann, T. Pfeiffer, M. Huebner, A.W. Bett, J. Becker, and J. Leuthold, "Optically powered fibre networks," in *Proc. 13th Biannual International Microwave and Optoelectronics Conference (IMOC'09)*, Belèm, Parà, Brazil. Session 'Optical Devices, Systems and Related Topics I', Nov. 3–6, 2009 [invited].

[C6] W. Freude, **M. Röger**, M. Dreschmann, M. Huebner, A.W. Bett, J. Becker, and J. Leuthold, "An optically powered fibre network for heterogeneous subscribers," in *Proc. 11th Intern. Conf. on Transparent Optical Networks (ICTON'09)*, Ponta Delgada, Island of São Miguel, Portugal, Paper Mo.B2.1, Vol. 1, pp. 1–4, June 28 – July 2, 2009 [invited].

[C7] **M. Röger**, M. Hoh, J. Hehmann, T. Pfeiffer, M. Hübner, J. Becker, J. Leuthold, and W. Freude, "Low-energy demarcation device for monitoring FTTx networks," in *Proc. 10. VDE-ITG-Fachtagung Photonische Netze*, Deutsche Telekom Hochschule für Telekommunikation, Leipzig, May 2009.

[C8] G. Böttger, M. Dreschmann, **M. Röger**, M. Hübner, C. Klamouris, A.W. Bett, T. Kueng, J. Becker, W. Freude, and J. Leuthold, "Optically powered video camera link," in *Proc. Optical Society of America Annual Meeting (OSA'07)*, San Jose (CA) USA, September 16-20, 2007, paper SME3 [invited].

[C9] G. Böttger, M. Dreschmann, **M. Röger**, M. Hübner, C. Klamouris, A.W. Bett, T. Kueng, J. Becker, W. Freude, and J. Leuthold, "Optically powered video camera link," in *Proc. 33th European Conf. Opt. Commun. (ECOC'07)*, Berlin, Germany, September 16–20, 2007, paper Thu9.6.3.

Patents

[P1] **M. Röger**, W. Freude, J. Leuthold, D. Volkmer, C. Klamouris, " Rotor Blade, Power Plant, and Use," Pub.No.: WO 2010/136151 A2.

[P2] J. Leuthold, W. Freude, T. Pfeiffer, G. Böttger, **M. Röger**, “Demarcation Devices, Network and Method,“ Pub.No.: WO/2008/037462.

Supervised Students' Work

[S1] S. Huber, “A Single Atom Transistor as Energy-Efficient Binary Charge Monitor for a Capacitor,” Bachelor Thesis, Institut für Photonik und Quantenelektronik (IPQ), Karlsruhe Institute of Technology, November 2011.

[S2] B. Hiba, “Weiterentwicklung eines Demarkationsgeräts zur Überwachung optischer Zugangsnetze (FTTH),” Diploma Thesis, Alcatel Lucent Deutschland, Bell Laboratories, Institut für Photonik und Quantenelektronik (IPQ), Karlsruhe Institute of Technology, August 2011.

[S3] M. Wippler, “Energieeffiziente Plattform für ein optisch versorgtes Sensornetzwerk,” Studienarbeit, Institut für Photonik und Quantenelektronik (IPQ), Karlsruhe Institute of Technology, July 2011.

[S4] E. Tietz, “Einsatz eines Ein-Atom-Transistor in einer Standby-Schaltung,” Bachelor Thesis, Institut für Photonik und Quantenelektronik (IPQ), Karlsruhe Institute of Technology, April 2011.

[S5] M. Hoh, “New Protocol for an Optically Powered Heterogeneous Sensor Network,” Studienarbeit, Institut für Photonik und Quantenelektronik (IPQ), Karlsruhe Institute of Technology, November 2010.

[S6] B. Hiba, “Optische Versorgung eines Demarkationsgerät,” Studienarbeit, Institut für Photonik und Quantenelektronik (IPQ), Karlsruhe Institute of Technology, November 2010.

[S7] A. Kleff, “Optisch versorgtes Netzwerk – Basisstation,” Bachelor Thesis, Institut für Photonik und Quantenelektronik (IPQ), Karlsruhe Institute of Technology, October 2010.

[S8] F. Boes, “Optisch versorgtes Netzwerk – Sensoreinheit,” Bachelor Thesis, Institut für Photonik und Quantenelektronik (IPQ), Karlsruhe Institute of Technology, October 2010.

[S9] M. Reichel, “Low-cost couplers for optical power and data transmission,” Studienarbeit, Institut für Photonik und Quantenelektronik (IPQ), Karlsruhe Institute of Technology, November 2009.

[S10] M. Dreschmann, “Rekonfigurierbare Hardware f¨ur Ultra-Low Power Over Fiber Anwendungen,” Diploma Thesis, Institut für Technik der Informationsverarbeitung (ITIV), Karlsruhe Institute of Technology, May 2009.

Karlsruhe Series in Photonics & Communications
KIT, Institute of Photonics and Quantum Electronics (IPQ)
(ISSN 1865-1100)

Die Bände sind unter www.ksp.kit.edu als PDF frei verfügbar oder als Druckausgabe bestellbar.

Band 1 Christian Koos
Nanophotonic Devices for Linear and Nonlinear Optical Signal Processing. 2008
ISBN 978-3-86644-178-1

Band 2 Ayan Maitra
Nonlinear Resonators for All-Optical Signal Processing. 2009
ISBN 978-3-86644-150-7

Band 3 Jin Wang
Pattern Effect Mitigation Techniques for All-Optical Wavelength Converters Based on Semiconductor Optical Amplifiers. 2008
ISBN 978-3-86644-276-4

Band 4 Jan-Michael Brosi
Slow-Light Photonic Crystal Devices for High-Speed Optical Signal Processing. 2009
ISBN 978-3-86644-313-6

Band 5 Christoph Dyroff
Tunable Diode-Laser Absorption Spectroscopy for Trace-Gas Measurements with High Sensitivity and Low Drift. 2009
ISBN 978-3-86644-328-0

Band 6 Philipp Vorreau
An Optical Grooming Switch for High-Speed Traffic Aggregation in Time, Space and Wavelength. 2010
ISBN 978-3-86644-502-4

Band 7 Thomas Vallaitis
Ultrafast Nonlinear Silicon Waveguides and Quantum Dot Semiconductor Optical Amplifiers. 2011
ISBN 978-3-86644-748-6

Band 8 René Bonk
Linear and Nonlinear Semiconductor Optical Amplifiers for Next-Generation Optical Networks. 2013
ISBN 978-3-86644-956-5

Karlsruhe Series in Photonics & Communications
KIT, Institute of Photonics and Quantum Electronics (IPQ)
(ISSN 1865-1100)

Band 9 David Hillerkuss
Single-Laser Multi-Terabit/s Systems. 2013
ISBN 978-3-86644-991-6

Band 10 Moritz Röger
Optically Powered Highly Energy-efficient Sensor Networks. 2013
ISBN 978-3-86644-972-5